Seeds of Destruction

F. William Engdahl

Seeds of Destruction

The Hidden Agenda
of Genetic Manipulation

Global Research

Seeds of Destruction. The Hidden Agenda of Genetic Manipulation,
by F. William Engdahl — First Edition

© F. William Engdahl, Global Research, Centre for Research on Globalization.
All rights reserved, 2007.

Global Research is a division of the Centre for Research on Globalization (CRG),
11, rue Notre-Dame Ouest, P.O. Box 55019, Montréal, Québec, H2Y 4A7, Canada.

For more information, contact the publisher at the above address or by email at
our website at **http://www.globalresearch.ca**

The views expressed herein are the sole responsibility of the author and do not
necessarily reflect those of the Centre for Research on Globalization. The publisher
will not be held liable for the material contained in this book or any statements,
omissions or inaccuracies pertaining thereto.

FIRST EDITION

Cover graphics by Nicolas Calvé and Sarah Choukah, © Global Research 2007

Printed and bound in Canada.
Printed on chlorine-free 100% post-consumer recycled paper.

ISBN 978-0-9737147-2-2

Legal Deposit:
Bibliothèque et Archives nationales du Québec
Library and Archives Canada

Library and Archives Canada Cataloguing in Publication

Engdahl, F. William, 1944-

    Seeds of destruction: the hidden agenda of genetic manipulation /
    F. William Engdahl.
    Includes bibliographical references and index.
    ISBN 978-0-9737147-2-2

1. Plant biotechnology–Political aspects–United States. 2. Plant genetic engineer-
ing–Political aspects–United States. 3. Transgenic plants–Political aspects–
United States. 4. Agricultural biotechnology–Political aspects–United States.
I. Centre for Research on Globalization. II. Title.

SB106.B56E54 2007   631.5'233   C2007-905814-0

I dedicate this book to Gottfried Gloeckner,
farmer, friend, self-taught scientist,
whose personal courage in the face of untold pressure
may have saved more lives than he or we shall ever know.

Also by F. William Engdahl

A Century of War:
Anglo-American Oil Politics and the New World Order

TABLE OF CONTENTS

Introduction

"We have about 50% of the world's wealth but only
6.3% of its population. This disparity is particularly
great as between ourselves and the peoples of Asia. In
this situation, we cannot fail to be the object of envy
and resentment. Our real task in the coming period is
to devise a pattern of relationships which will permit
us to maintain this position of disparity without positive detriment to our national security. To do so, we will
have to dispense with all sentimentality and day dream-
ing; and our attention will have to be concentrated
everywhere on our immediate national objectives. We
need not deceive ourselves that we can afford today the
luxury of altruism and world-benefaction."

George Kennan,
US State Department senior planning official, 1948

This book is about a project undertaken by a small socio-political
elite, centered, after the Second World War, not in London,
but in Washington. It is the untold story of how this self-anointed
elite set out, in Kennan's words, to "maintain this position of disparity." It is the story of how a tiny few dominated the resources and
levers of power in the postwar world.

It's above all a history of the evolution of power in the control
of a select few, in which even science was put in the service of that
minority. As Kennan recommended in his 1948 internal memorandum, they pursued their policy relentlessly, and without the
"luxury of altruism and world-benefaction."

Yet, unlike their predecessors within leading circles of the British Empire, this emerging American elite, who proclaimed proudly at war's end the dawn of their American Century, were masterful in their use of the rhetoric of altruism and world-benefaction to advance their goals. Their American Century paraded as a softer empire, a "kinder, gentler" one in which, under the banner of colonial liberation, freedom, democracy and economic development, those elite circles built a network of power the likes of which the world had not seen since the time of Alexander the Great some three centuries before Christ—a global empire unified under the military control of a sole superpower, able to decide on a whim, the fate of entire nations.

This book is the sequel to a first volume, *A Century of War: Anglo-American Oil Politics and the New World Order*. It traces a second thin red line of power. This one is about the control over the very basis of human survival, our daily provision of bread. The man who served the interests of the postwar American-based elite during the 1970's, and came to symbolize its raw *realpolitik*, was Secretary of State Henry Kissinger. Sometime in the mid-1970's, Kissinger, a life-long practitioner of "Balance of Power" geopolitics and a man with more than a fair share of conspiracies under his belt, allegedly declared his blueprint for world domination: "*Control the oil and you control nations. Control the food, and you control the people.*"

The strategic goal to control global food security had its roots decades earlier, well before the outbreak of war in the late 1930's. It was funded, often with little notice, by select private foundations, which had been created to preserve the wealth and power of a handful of American families.

Originally the families centered their wealth and power in New York and along the East Coast of the United States, from Boston to New York to Philadelphia and Washington D.C. For that reason, popular media accounts often referred to them, sometimes with derision but more often with praise, as the East Coast Establishment.

The center of gravity of American power shifted in the decades following the War. The East Coast Establishment was overshadowed by new centers of power which evolved from Seattle to

Southern California on the Pacific Coast, as well as in Houston, Las Vegas, Atlanta and Miami, just as the tentacles of American power spread to Asia and Japan, and south, to the nations of Latin America.

In the several decades before and immediately following World War II, one family came to symbolize the hubris and arrogance of this emerging American Century more than any other. And the vast fortune of that family had been built on the blood of many wars, and on their control of a new "black gold," oil.

What was unusual about this family was that early on in the building of their fortune, the patriarchs and advisors they cultivated to safeguard their wealth decided to expand their influence over many very different fields. They sought control not merely over oil, the emerging new energy source for world economic advance. They also expanded their influence over the education of youth, medicine and psychology, foreign policy of the United States, and, significant for our story, over the very science of life itself, biology, and its applications in the world of plants and agriculture.

For the most part, their work passed unnoticed by the larger population, especially in the United States. Few Americans were aware how their lives were being subtly, and sometimes not so sub tly, influenced by one or another project financed by the immense wealth of this family.

In the course of researching for this book, a work nominally on the subject of genetically modified organisms or GMO, it soon became clear that the history of GMO was inseparable from the political history of this one very powerful family, the Rockefeller family, and the four brothers—David, Nelson, Laurance and John D. III—who, in the three decades following American victory in World War II, the dawn of the much-heralded American Century, shaped the evolution of power George Kennan referred to in 1948.

In actual fact, the story of GMO is that of the evolution of power in the hands of an elite, determined at all costs to bring the entire world under their sway.

Three decades ago, that power was based around the Rockefeller family. Today, three of the four brothers are long-since deceased, several under peculiar circumstances. However, as was their will, their project of global domination—"full spectrum dominance" as the Pentagon later called it—had spread, often through a rhetoric of "democracy," and was aided from time to time by the raw military power of that empire when deemed necessary. Their project evolved to the point where one small power group, nominally headquartered in Washington in the early years of the new century, stood determined to control future and present life on this planet to a degree never before dreamed of.

The story of the genetic engineering and patenting of plants and other living organisms cannot be understood without looking at the history of the global spread of American power in the decades following World War II. George Kennan, Henry Luce, Averell Harriman and, above all, the four Rockefeller brothers, created the very concept of multinational "agribusiness". They financed the "Green Revolution" in the agriculture sector of developing countries in order, among other things, to create new markets for petro-chemical fertilizers and petroleum products, as well as to expand dependency on energy products. Their actions are an inseparable part of the story of genetically modified crops today.

By the early years of the new century, it was clear that no more than four giant chemical multinational companies had emerged as global players in the game to control patents on the very basic food products that most people in the world depend on for their daily nutrition—corn, soybeans, rice, wheat, even vegetables and fruits and cotton—as well as new strains of disease-resistant poultry, genetically-modified to allegedly resist the deadly H5N1 Bird Flu virus, or even gene altered pigs and cattle. Three of the four private companies had decades-long ties to Pentagon chemical warfare research. The fourth, nominally Swiss, was in reality Anglo-dominated. As with oil, so was GMO agribusiness very much an Anglo-American global project.

In May 2003, before the dust from the relentless US bombing and destruction of Baghdad had cleared, the President of the United States chose to make GMO a strategic issue, a priority in his post-war US foreign policy. The stubborn resistance of the world's second largest agricultural producer, the European Union, stood as a formidable barrier to the global success of the GMO Project. As long as Germany, France, Austria, Greece and other countries of the European Union steadfastly refused to permit GMO planting for health and scientific reasons, the rest of the world's nations would remain skeptical and hesitant. By early 2006, the World Trade Organization (WTO) had forced open the door of the European Union to the mass proliferation of GMO. It appeared that global success was near at hand for the GMO Project.

In the wake of the US and British military occupation of Iraq, Washington proceeded to bring the agriculture of Iraq under the domain of patented genetically-engineered seeds, initially supplied through the generosity of the US State Department and Department of Agriculture.

The first mass experiment with GMO crops, however, took place back in the early 1990's in a country whose elite had long since been corrupted by the Rockefeller family and associated New York banks: Argentina.

The following pages trace the spread and proliferation of GMO, often through political coercion, governmental pressure, fraud, lies, and even murder. If it reads often like a crime story, that should not be surprising. The crime being perpetrated in the name of agricultural efficiency, environmental friendliness and solving the world hunger problem, carries stakes which are vastly more important to this small elite. Their actions are not solely for money or for profit. After all, these powerful private families decide who controls the Federal Reserve, the Bank of England, the Bank of Japan and even the European Central Bank. Money is in their hands to destroy or create.

Their aim is rather, the ultimate control over future life on this planet, a supremacy earlier dictators and despots only ever dreamt of. Left unchecked, the present group behind the GMO Project is between one and two decades away from total dominance of the planet's food capacities. This aspect of the GMO story needs telling. I therefore invite the reader to a careful reading and independent verification or reasoned refutation of what follows.

PART I
The Political Beginnings

CHAPTER 1
Washington Launches
the GMO Revolution

Early GMO Research

The issue of biotechnology and genetic-modification of plants and other life forms first emerged from research labs in the United States in the late 1970's. During the 1980's, the Reagan Administration acted in key areas of economic policy in ways which echoed the radical policies of the President's close ally, British Prime Minister Margaret Thatcher. There was a special relationship between the two, as both were deeply committed advocates of radical free market policies and less government involvement, combining to give the private sector free reign.

In one domain, however, that of the emerging field of genetic engineering which developed, some years before, out of DNA (Deoxyribonucleic Acid) and RNA (Ribonucleic Acid) research, Reagan's Administration was determined to take a back seat to no-one in seeing to it that America was Number One.

A curious aspect of the regulatory history of GMO foods and genetically-engineered products in the United States was that, beginning in the Reagan era, the government showed extreme partisanship in favour of the biotech agribusiness industry. The very

US Government agencies entrusted with the mandate to safeguard the health and safety of the overall population were becoming dangerously biased.

Some years before, the first commercial GMO product hit the market in the US, the Reagan Administration had been moving quietly to open its doors wide to Monsanto and other private companies which were developing gene-manipulated products. The key actor within the Reagan Administration on decisions pertaining to the new field of genetically modified products was former head of the CIA, Vice President George Herbert Walker Bush—who would himself soon be President, and father of the later President, George W. Bush.

By the early 1980's, numerous agribusiness corporations were in a gold rush frenzy to develop GMO plants, livestock and GMO-based animal drugs. There was no regulatory system in place to control the development, risks and sale of the products. The agribusiness companies wanted to keep it that way.

The Reagan-Bush Administration was partly driven by an ideological agenda of imposing deregulation, reducing Government supervision in every facet of daily life. Food safety was no exception. Rather to the contrary, and even if that meant the general population could become guinea pigs for entirely untested new health risks.

The Fraud of "Substantial Equivalence"
In 1986, Vice President Bush hosted a group of executives from a giant chemical company, Monsanto Corporation of St. Louis, Missouri, for a special White House strategy meeting. The purpose of the unpublicized meeting, according to former US Department of Agriculture official, Claire Hope Cummings, was to discuss the "deregulation" of the emerging biotech industry. Monsanto had had a long history of involvement with the US Government and even with Bush's CIA. It had developed the deadly herbicide, Agent Orange, for defoliation of jungle areas in Vietnam during the 1960's. It also had a long record of fraud, cover-up and bribery.

When he finally became President in 1988, Bush and his Vice President Dan Quayle moved swiftly to implement an agenda giving unregulated free-rein to Monsanto and other major GMO companies. Bush decided it was time to make public the regulatory framework which he had negotiated a few years earlier behind closed doors.

Vice President Quayle, as head of Bush's Council on Competitiveness, announced that "biotech products will receive the same oversight as other products," and "not be hampered by unnecessary regulation."[1] On May 26, 1992, Vice President Dan Quayle proclaimed the Bush administration's new policy on bio-engineered food.

"The reforms we announce today will speed up and simplify the process of bringing better agricultural products, developed through biotech, to consumers, food processors and farmers," Mr. Quayle told executives and reporters. "We will ensure that biotech products will receive the same oversight as other products, instead of being hampered by unnecessary regulation."[2] Pandora's Box had been opened by the Bush-Quayle Administration.

Indeed, not one single new regulatory law governing biotech or GMO products was passed then or later, despite repeated efforts by concerned Congressmen that such laws were urgently needed to regulate unknown risks and possible health dangers from the genetic engineering of foods.

The framework that Bush put in place was simple. In line with the expressed wishes of the biotech industry, the US Government would regard genetic engineering of plants and foods or animals as merely a simple extension of traditional animal or plant breeding.

Further clearing the path for Monsanto and company, the Bush Administration decided that traditional agencies, such as the US Department of Agriculture, the EPA, the Food and Drug Administration (FDA) and the National Institutes of Health (NIH), were competent to evaluate the risks of GMO products.[3] They determined that no special agency was needed to oversee the revolutionary new field. Furthermore, the responsibilities for the four different agencies were kept intentionally vague.

That vagueness ensured overlap and regulatory confusion, allowing Monsanto and the other GMO operators maximum leeway to introduce their new genetically engineered crops. Yet, to the outside world, it appeared that the new GMO products were being carefully screened. The general public naturally assumed that the Food and Drug Administration or the National Institutes of Health were concerned about their well-being.

Despite serious warnings from research scientists about the dangers of recombinant DNA research and biotechnology work with viruses, the US Government opted for a system in which the industry and private scientific laboratories would "voluntarily" police themselves in the new field of genetically engineered plants and animals.

There were repeated warnings from senior US government scientists of the potential dangers to the Bush-Quayle "no regulation" decision. Dr. Louis J. Pribyl, of the Food & Drug Administration was one of 17 government scientists working on a policy for genetically engineered food at the time. Pribyl knew from studies that toxins could be unintentionally created when new genes were introduced into a plant's cells. Pribyl wrote a heated warning memo to the FDA Chief Scientist declaring, "This is the industry's pet idea, namely that there are no unintended effects. ... But time and time again, there is no data to back up their contention."

Other Government scientists concluded there was "ample scientific justification" to require tests and a government review of each genetically engineered food before it was sold. "The possibility of unexpected, accidental changes in genetically engineered plants justifies a limited traditional toxicological study," they declared.[4] Their voices went unheeded by the Bush Administration. They had cut their deal with Monsanto and the emerging biotech agribusiness industry.

At that early stage, few paid any attention to the enormous implications of genetic engineering on such a mass scale, outside a small circle of scientists being financed by the largesse of a handful of foundations. And no foundation was more important in the financing of this emerging sector of biotechnology than the Rockefeller Foundation in New York.

By 1992, President George H.W. Bush was ready to open the Pandora's Box of GMO. In an Executive Order, the President made the ruling that GMO plants and foods were "substantially equivalent" to ordinary plants of the same variety, such as ordinary corn, soybeans, rice or cotton.[5]

The doctrine of "substantial equivalence" was the lynchpin of the whole GMO revolution. It meant that a GMO crop could be considered to be the same as a conventional crop, merely because GMO corn looked like ordinary corn or GM rice or soybean, and even tasted more or less like conventional corn, and because in its chemical composition and nutritional value, it was "substantially" the same as the natural plant.

That determination that GMO plants were to be treated as "substantially equivalent" ignored the qualitative internal alteration required to genetically engineer the particular crop. As serious scientists pointed out, the very concept of "substantial equivalence" was itself pseudo-scientific. The doctrine of "substantial equivalence" had been created primarily to provide an excuse for not requiring biochemical or toxological tests.

Because of the Bush Administration's "substantial equivalence" ruling, no special regulatory measures would be required for genetically engineered varieties.

Substantial equivalence was a phrase which delighted the agribusiness companies. That wasn't surprising, for Monsanto and the others had created it. Its premise was deceitful, as Bush's science advisers well knew.

Genetic modification of a plant or organism involved taking foreign genes and adding them to a plant such as cotton or soybeans to alter their genetic makeup in ways not possible through ordinary plant reproduction. Often the introduction was made by a gene "cannon" literally blasting a plant with a foreign bacteria or DNA segment to alter its genetic character. In agricultural varieties, hybridization and selective breeding had resulted in crops adapted to specific production conditions and regional demands.

Genetic engineering differed from traditional methods of plant and animal breeding in very important respects. Genes from one organism could be extracted and recombined with those of another (using recombinant DNA, or rDNA, technology) without either organism having to be of the same species. Second, removing the requirement for species reproductive compatibility, new genetic combinations could be produced in a highly accelerated way. The fateful Pandora's Box had indeed been opened. The fictional horrors of the "Andromeda Strain," the unleashing of a biological catastrophe, was no longer the stuff of science fiction. The danger was real, and no one seemed to be overtly concerned.

Genetic engineering introduced a foreign organism into a plant in a process that was imprecise and unpredictable. The engineered products were no more "substantially equivalent" to the original than a tiny car hiding a Ferrari engine would be to a Fiat.

Ironically, while companies such as Monsanto argued for "substantial equivalence," they also claimed patent rights for their genetically modified plants on the argument that their genetic engineering had created substantially new plants whose uniqueness had to be protected by exclusive patent protection. They saw no problem in having their cake and eating it too.

With the Bush Administration 1992 ruling, that was to be upheld by every successive Administration, the US Government treated GMO or bio-engineered foods as "natural food additives," therefore not subjecting them to any special testing. If it wasn't necessary to test normal corn to see if it was healthy to eat, so went the argument, why should anyone have to bother to test the "substantially equivalent" GM corn, soybean, or GM milk hormones produced by Monsanto and the other agribusiness companies?

In most cases, the Government regulatory agencies simply took the data provided to them by the GMO companies themselves in order to judge that a new product was fine. The US Government agencies never ruled against the gene giants.

"Nature's Most Perfect Food…"

The first mass-marketed GMO food was milk containing a recombinant Bovine Growth Hormone, known as rBGH. This was a genetic manipulation patented by Monsanto. The FDA declared the genetically-engineered milk safe for human consumption before crucial information on how the GM milk might affect human health was available, diligently holding up to the doctrine of sub-stantial equivalence.

The rBGH hormone constituted a huge temptation for strug-gling dairy farmers. Monsanto claimed that if injected regularly with rBGH, which it sold under the trade name Posilac, cows would typically produce up to thirty percent more milk. For the strug-gling farmer, a thirty percent jump in output per cow was aston-ishing and virtually irresistible. Monsanto advertised that farmers should "leave no cow untreated." One state agriculture commis-sioner termed rBGH "crack for cows" because of its extraordinary stimulating effects on milk output.[6]

Monsanto's new Posilac rBGH hormone not only stimulated the cow to produce more milk. In the process it stimulated production of another hormone, IGF-1, which regulated the cow's metabolism, in effect, stimulating the cell division within the animal and hin-dering cell death. This is where problems began to appear.

Various independent scientists spoke out, warning that Monsanto's rBGH hormone increased the levels of insulin-like growth factors, and had a possible link to cancer. One of the most vocal scientists on the matter was Dr. Samuel Epstein, from the University of Illinois's School of Public Health. Epstein, a recognized authority on carcinogens, warned of a growing body of scientific evidence that the Insulin-like Growth Factor (IGF-1), was linked to the creation of human cancers, cancers which might not appear for years after initial exposure.[7]

Not surprisingly, hormone stimulation that got cows to pump 30% more milk had other effects. Farmers began to report their cows burned out by as much as two years sooner, and that many cows had serious hoof or udder infections as a by-product of the rBGH hormone treatment, meaning that some of them could not walk. In turn, the cows had to be injected with more antibiotics to treat those effects.

The FDA countered the growing criticism by using data pro-vided by Monsanto, which, not surprisingly, severely criticized the independent scientists. With Monsanto's chief rBGH scientist, Dr. Robert Collier, with tongue firmly in cheek, retorted that, "In fact the FDA has commented several times on this issue. ... They have publicly restated human safety confidence ... this is not something knowledgeable people have concerns about."[8] That was hardly reassuring for anyone aware of the relationship between Monsanto and the FDA leadership.

In 1991, a scientist at the University of Vermont leaked to the press that there was evidence of severe health problems affecting rBGH-treated cows, including mastitis, an inflammation of the udder, and deformed births. Monsanto had spent more than half a million dollars to fund the University of Vermont test trials of rBGH. The chief scientist of the project, in direct opposition to his alarmed researcher, had made numerous public statements asserting that rBGH cows had no abnormal levels of health problems compared with regular cows. The unexpected leak from the upstart whistleblower was embarrassing for both Monsanto and the University receiving Monsanto research dollars, to say the least.[9]

The US General Accounting Office, an investigative arm of the US Congress, was called in to investigate the allegations. Both the University of Vermont and Monsanto refused to cooperate with the GAO, which was finally forced to give up the investigation with no results. Only years later did the University finally release the data, which indeed showed the negative health effects of rBGH. By then, however, it was too late.

In 1991, the Food and Drug Administration created the new position of Deputy Commissioner for Policy to oversee agency policy on GMO foods. The agency named Michael R. Taylor to be its first head. Taylor came to the job as a Washington lawyer. But not just any old garden variety of Washington lawyers. As a food and drug law specialist with the Washington power firm, King & Spalding, Taylor had previously successfully represented Monsanto and other biotech companies in regulatory cases.[10]

Monsanto's chief scientist, Margaret Miller, also assumed a top post in the FDA as Deputy Director of Human Food Safety at the beginning of the 1990's. In this position, Dr. Miller, without an explanation, raised the FDA standard by 100 times for the permissible level of antibiotics that farmers could put into milk. She single-handedly cleared the way for a booming business for Monsanto's rBGH hormone. A cozy club was emerging between private biotech companies and the government agencies that should be regulating them. It was a club more than a little fraught with potential conflict of interest.[11]

As one of its top officials, Taylor helped the FDA draw up guidelines to decide whether GMO foods should be labeled. His decision was not to label GMO foods.

At the same time, again under Taylor's guiding hand, the FDA ruled that risk-assessment data, such as data on birth defects in cattle or even possible symptoms in humans arising from consumption of GMO foods, could be withheld from the public as "confidential business information."

Were it to leak out that Monsanto, Dow or other biotech companies were creating grotesque deformities in animals fed GMO foods, it might be detrimental for the stock price of the company, and that would damage the full flowering of private enterprise.

This, at least, seemed the logic behind the perverse kind of "Shareholder value ueber Alles." As FDA Biotechnology Coordinator James Maryansky remarked, "The FDA would not require things to be on the label just because a consumer might want to know them."[12]

A lawyer for Monsanto, Michael R. Taylor, had been placed in charge of GMO food policy within the government's principal food safety body. As a suitable postscript, honoring the adage, "we take care of our friends," Monsanto rewarded the diligent public servant by appointing Michael Taylor to be Vice President of Monsanto for Public Policy after he left the FDA.[13]

FDA and Monsanto Milk the Public

By 1994, after a suitable amount of time had elapsed, the FDA approved the sale of rBGH milk to the public. Under the FDA rules, of course, it was unlabeled, so the consumer could avoid undue anxiety about giving himself or his children exposure to cancerous agents or other surprises. He would never know. When Monsanto's Posilac caused leukemia and tumors in rats, the US Pure Food and Drug Act was rewritten to allow a product that caused cancer in laboratory animals to be marketed for human consumption without a warning label. It was as simple as that.

Though Monsanto claimed that its rBGH was one of the most thoroughly examined drugs in US history, rBGH was never tested in the long-term for (chronic) human health effects. A generally accepted principle in science holds that two years of testing is the minimal time for long-term health studies. rBGH was tested for only 90 days on 30 rats. The short-term rat study was submitted by Monanto to the FDA but was never published. The FDA refused to allow anyone outside the administration to review the raw data from this study, saying that publication would "irreparably harm" Monsanto. Monsanto has continued to refuse to allow open scientific peer review of the 90 day study. This linchpin study of cancer and BGH has never been subjected to scrutiny by the scientific community.[14]

Not content to feed GMO milk exclusively to its own unwary population, the US Government exerted strong pressure on Mexico and Canada also to approve rBGH, as part of an effort to expand Monsanto's rBGH market globally.

However, the FDA-Monsanto campaign got a nasty setback in January 1999, when the Canadian counterpart to the FDA, Health Canada, broke ranks with the US and issued a formal "notice of non-compliance" disapproving future Canadian sales of rBGH, sometimes also called rBST or recombinant Bovine Somatotropin.

The action followed strong pressure from the Canadian Veterinary Medical Association and the Royal College of Physicians, which presented evidence of the adverse effects of rBGH milk, including evidence of lameness and reproductive problems. Monsanto had been very eager to break into the Canadian market with its rBGH, even to the point, according to a Canadian CBC television report, that a Monsanto official tried to bribe a Canadian health official sitting on the Government review committee with an offer of $1-2 million, to secure rBGH approval in Canada without further studies. The insulted official reportedly asked, "Is that a bribe?" and the meeting ended.[15]

Moreover, a special European Commission independent committee of recognized experts concluded that rBGH, as reported in Canadian findings, not only posed the above-named dangers, but also major risks especially of breast and prostate cancer in humans.

In August 1999, the United Nations Food Safety Agency, the Codex Alimentarius Commission, ruled unanimously in favor of a 1993 European Union moratorium on the introduction of Monsanto's rBGH milk. Monsanto's rBGH was thus banned from the EU.[16]

This setback was not to daunt the persistent bureaucrats at the FDA, or their friends at Monsanto. Since GMO labeling had been forbidden by the FDA, Americans were blissfully unaware of the dangers of drinking the milk they were encouraged to consume for better health. "Nature's most perfect food" was the dairy industry's slogan for milk. With regard to reporting the UN decision and the negative Canadian conclusions, the US media were respectfully quiet. Americans were simply told that the EU was trying to hurt American cattle farmers by refusing imports of hormone-fed US beef.

One concerned FDA scientist who refused to sit by idly was FDA Veterinarian Dr. Richard Burroughs, who was responsible, from 1979 until 1989, for reviewing animal drugs such as rBGH. From 1985 until the year he was fired, Burroughs headed the FDA's review of Monsanto's rBGH, thus being directly involved in the evaluation process for almost five years. Burroughs wrote the original protocols for animal safety studies and reviewed the data submitted by rBGH developers from their own safety studies.

In a 1991 article in *Eating Well* magazine, Burroughs described a change in the FDA beginning in the mid-1980s. Burroughs was faced with corporate representatives who wanted the FDA to ease strict safety testing protocols. He reported seeing corporations dropping off sick cows from rBGH test trials and then manipulating data in such ways as to make health and safety problems "disappear."[17]

Burroughs challenged the agency's lenience and its changing role from guardian of public health to protector of corporate profits. He criticized the FDA and its handling of rBGH in statements to Congressional investigators, in testimonies to state legislatures, and in declarations to the press. Within the FDA, he rejected a number of corporate-sponsored safety studies, calling them insufficient. Finally, in November 1989, he was fired for "incompetence."

The FDA failed to act on the evidence that rBGH was not safe. In fact, the agency promoted the Monsanto Corporation's product before and after the drug's approval. Dr. Michael Hansen of Consumers Union noted that the FDA acted as an rBGH advocate by issuing news releases promoting rBGH, making public statements praising the drug, and writing promotional pieces about rBGH in the agency's publication, *FDA Consumer*.[18]

In April 1998, two enterprising award-winning television journalists at Fox TV, an influential US network owned by Rupert Murdoch, put together the remarkable story of the rBGH scandal including its serious health effects. Upon pressure by Monsanto, Fox killed the story and fired Jane Akre and her husband Steve. In an August 2000 Florida state court trial, the two won a jury award of $425,000 damages and the Court found that Fox "acted intentionally and deliberately to falsify or distort the plaintiffs' news reporting on rBGH."[19]

With their ample financial resources, Fox Television and Monsanto took the case to a higher court on appeal and got the decision reversed on a legal technicality. The FDA kept silent. Monsanto continued to market rBGH milk unabated. As one former US Department of Agriculture official stated, the guiding regulatory percent for genetically modified foods was, "don't tell, don't ask," which meant, "If the industry does not tell government what it knows about its GMOs, the government does not ask."[20] That was little reassurance for the health and safety concerns of the population. Few ever realized it however, as on the surface it appeared that the FDA and other relevant agencies were guarding their health interests in the new area of GMO foods.

In January 2004, after FDA inspectors broke their silence by declaring having found unacceptable levels of contamination in rBGH, Monsanto finally announced it would reduce the supply of Posilac by 50%. Many thought Monsanto would quietly discontinue production of the dangerous hormone. Not easily deterred by anything, least of all evidence of danger to human health, Monsanto announced a year later that they planned to increase the supply of Posilac again, initially to 70% of its peak level. They had come under enormous pressure not only from citizens concerned about health consequences, but also from farmers who realized that the 30% rise in national milk output from dairy herds had only served to create an even larger glut of unsold milk in a nation already in surplus. It had also triggered collapsing milk prices.

By then, Monsanto had moved on to corner the global market in seeds for the most important staples in the human and animal diets.

Monsanto's Cozy Government Relations

The relation between the US Government and giant GMO seed producers such as Monsanto, DuPont or Dow AgriSciences was not accidental. The Government encouraged development of unregulated GMO crops as a strategic priority, as noted, since the early years of the Reagan Presidency, long before it was at all clear whether such engineering of nature was at all desirable. It was one thing for a government to support long-term laboratory research through science grants. It was quite another thing to open the market's floodgates to untested, risky new procedures which had the potential to affect the basic food supply of the country and of the entire planet.

Washington was becoming infamous for what some called "revolving door government." The latter referred to the common practice of major corporations to hire senior government officials directly from government service into top corporate posts where their government influence and connections would benefit the company. Similarly, the practice worked in reverse, where top corporate persons got picked for prime government jobs where they could promote the corporation's private agenda inside the government.

Few companies were more masterful at this game of the revolving door than Monsanto. That corporation was a major contributor to both Republican and Democratic national candidates. During the controversy over the labeling of Monsanto's rBGH milk, the 12 members of the Dairy Subcommittee of the House Agricultural committee were no strangers to Monsanto's campaign largesse. They had won a total of $711,000 in Monsanto campaign finance. It is not possible to prove that this fact influenced the Committee's decision. However, it evidently did not hurt Monsanto's case. The Committee killed the proposed labeling law.

Monsanto had a special skill in placing its key people in relevant Government posts. George W. Bush's Agriculture Secretary, Ann Veneman, came to Washington in 2001 from a job as director of Calgene, a biotech company which became a Monsanto subsidiary. Defense Secretary Donald Rumsfeld had been CEO of Monsanto subsidiary G.D. Searle, producer of GMO-based artificial sweetener and carcinogen, Aspartame. Rumsfeld had also been Chairman of California biotech company Gilead Science, which held the patent on Tamiflu.

Former US Trade Representative and lawyer to Bill Clinton, Mickey Kantor, left Government to take a seat on the Board of Monsanto. Monsanto also had on its board William D. Ruckelshaus, former head of the Environmental Protection Agency (EPA) under Presidents Nixon and Reagan. Michael A. Friedman, M.D., senior vice-president of clinical affairs for Monsanto's pharmaceutical division G.D. Searle, was once acting director of the FDA. Marcia Hale, Monsanto's director of UK government affairs, was formerly an assistant to President Clinton for intergovernmental affairs. Linda J. Fisher, Monsanto vice president of public affairs, was once administrator of EPA's Office of Prevention, Pesticides, and Toxic Substances. Monsanto legal adviser, Jack Watson, was chief of White House staff in the Carter Administration.

This pattern of revolving door conflicts of interest between top officials of government agencies responsible for food policy and their corporate sponsors, such as Monsanto, Dow, DuPont and the other agribusiness and biotech players, had been in place at least since the time of the Reagan Administration.

Unmistakable was the conclusion that the US Government was an essential catalyst for the "gene revolution" of GMO-altered food crops and their proliferation worldwide. In this they acted in concert with the corporate giant agrichemical firms such as Monsanto, Dow and DuPont, as if public and private interests were the same.

What could explain the extraordinary backing of no fewer than four US Presidents for the GMO agrichemical industry? What could explain why Bill Clinton put the very authority of his office on the line to demand that the British Prime Minister silence a critic of the genetic manipulation of plants?

What could explain the extraordinary ability of firms such as Monsanto to get their way among government officials regardless of overwhelming evidence of potential health damage to the population? What could cause four Presidents to expose the health of their nation and the entire world to untold risks, against the warnings of countless scientists and even government officials responsible for public health regulation?

The answer to those questions was there for anyone willing to look. But it was an answer so shocking that few dared to examine it. A press conference in late 1999 gave a hint as to the powerful interests standing behind public players. On October 4, 1999, Gordon Conway, the President of an influential private tax-exempt foundation based in New York, applauded the announcement by Monsanto that it agreed not to "commercialize" its controversial "terminator" seed genetics.[21]

The organization was the Rockefeller Foundation. It was no coincidence that the Rockefeller Foundation and Monsanto were talking about a global strategy for the genetic engineering of plants. The genetic revolution had been a Rockefeller Foundation project from the very beginning. Not only, as Conway reminded in his public remarks, had the Rockefeller Foundation spent more than $100 million for the advance of the GMO revolution. That project was part of a global strategy that had been in development for decades.

At the 1999 press conference, Conway declared, "The Rockefeller Foundation supports the Monsanto Company's decision not to commercialize sterile seed technologies, such as the one dubbed "the Terminator." He added, "We welcome this move as a first step toward ensuring that the fruits of plant biotechnology are made available to poor farmers in the developing world."[22]

Conway had gone to Monsanto some months before to warn the senior executives that they risked jeopardizing the entire GMO revolution and that a tactical retreat was needed to keep the broad project on track.[23]

Terminator seeds had been designed to prevent the germination of harvested grains as seeds, and had engendered strong opposition in many quarters. This technology would block farmers in developing from saving their own seed for re-sowing.[24]

The involvement of the Rockefeller Foundation in Monsanto's corporate policy was not by chance. It was part of a far more ambitious plan rooted in the crisis of the post-war dollar order which began in the era of the Vietnam War.

This technology would prevent farmers in developing countries from saving their own seed for re-sowing.[25]

The involvement of the Rockefeller Foundation in Monsanto's corporate policy was not by chance. It was part of a far more ambitious plan which began in the era of the Vietnam War. The GMO project required that scientists should serve their agribusiness patrons. The development of a research project in Scotland was intended to send a strong signal to biologists around the world as to what happens when the results of GMO research contradict the interests of Monsanto and other GMO producers.

Notes

1. Quoted in Kurt Eichenwald et al., "Biotechnology Food: From the Lab to a Debacle", *New York Times*, 25 January 2001.

2. *Ibid.*

3. Dr. Henry Miller, quoted in Eichenwald et al., *op.cit.* Miller, who was responsible for biotechnology issues at the Food and Drug Administration from 1979 to 1994, told the New York Times: "In this area, the U.S. government agencies have done exactly what big agribusiness has asked them to do and told them to do."

4. Eichenwald, *op. cit.*

5. Claire Hope Cummings, *Are GMOs Being Regulated or Not?*, 11 June 2003, in http://www.cropchoice.com/leadstry66f7.html?recid=1736. Cummings was a senior US Department of Agriculture official at the time.

6. Jeffrey Smith, *Got Hormones—The Controversial Milk Drug that Refuses to Die,* December 2004, http://www.responsibletechnology.org.

7. Robert P. Heaney, et al., "Dietary Changes Favorably Affect Bone Remodeling in Older Adults," *Journal of the American Dietetic Association*, vol. 99, no. 10, October 1999, pp. 1228-1233. Also, "Milk, Pregnancy, Cancer May Be Tied", *Reuters*, 10 September 2002.

8. Dr. Robert Collier, quoted in Jane Akre & Steve Wilson, from text of banned FOX TV documentary, "The Mystery in Your Milk," in http://www.mercola.com/2001/may/26/mystery_milk.htm.

9. Jennifer Ferrara, "Revolving Doors: Monsanto and the Regulators", *The Ecologist*, September/October 1998.

10. Michael R. Taylor, "Biography", in *Food Safety Research Consortium, Steering Committee*, in http://www.rff.org/fsrc/bios.htm.

11. Robert Cohen, *FDA Regulation Meant to Promote rBGH Milk Resulted in Antibiotic Resistance*, 5 May 2000, in http://www.psrast.org/bghsalmonella.htm.

12. James Maryansky, quoted in Julian Borger, "Why Americans are Happy to Swallow the GM Food Experiment", *The Guardian*, 20 February 1999.

13. Steven M. Druker, *Bio-deception: How the Food and Drug Administration is Misrepresenting the Facts about Risks of Genetically Engineered Foods...*, http://www.psrast.org/fdalawstmore.htm. Druker drafted the statement in May 1998 as part of a lawsuit against the FDA to demand mandatory testing and labeling of GMO food, both of which are not, as of 2007, done in the United States.

14. In his book, *Milk, the Deadly Poison*, Argus Press, Inglewood Cliffs, NJ, 1997, pp. 67-96, Robert Cohen describes his efforts to obtain a copy of this unpublished study from the FDA. Cohen filed a Freedom of Information Act request

for the study and it was denied; he appealed within the FDA and lost. He then filed a lawsuit in federal court and again, lost. The FDA and the courts agree that the public should never learn what happened to those rats fed BGH because it would "irreparably harm" Monsanto. Based on the scant information that has been published about the weight gains of the rats during the 90-day study, Cohen believes that many or perhaps all of the rats got cancer.

15. In November 1994, the Canadian Broadcasting Corporation (CBC) program The Fifth Estate televised a one-hour documentary reporting that Monsanto had tried to bribe Health Canada (Canada's equivalent to the FDA), offering to pay as much as two million dollars under the condition that Monsanto receive approval to market rBGH in Canada without being required to submit data from any further studies or trials. According to journalists who worked on the documentary, Monsanto tried to kill the show, arguing through its lawyers that CBC had maliciously rigged the interviews. But CBC stuck to its guns and ran the program.

16. PRNewswire, *Monsanto's Genetically Modified Milk Ruled Unsafe by the United Nations*, Chicago, 18 August 1999. John R. Luoma, "Pandora's Pantry", *Mother Jones*, January/February 2000.

17. Robert Cohen, *FDA Regulation Meant to Promote rBGH Milk Resulted in Antibiotic Resistance*, http://www.psrast.org, 5 May 2000.

18. *Ibid.*

19. RBGH Bulletin, *Hidden Danger in Your Milk?: Jury Verdict Overturned on Legal Technicality*, http://www.foxrBGHsuit.com, 2000.

20. The Agribusiness Examiner, *Kraft "Cheese?": Adulterated Food?—FDA: Don't Ask! Don't Tell!*, 7 May 2001, http://www.mindfully.org/Food/Kraft-Cheese-Adulterated.htm.

21. Dr. Gordon Conway in a speech to Directors of Monsanto, "The Rockefeller Foundation and Plant Biotechnology", 24 June 1999, in http://www.biotech-info.net/gordon_conway.html.

22. Rockefeller Foundation, Press Release, "*Terminator' Seed Sterility Technology Dropped*, 4 October 1999, in http://www.rockfound.org/.

23. John Vidal, "How Monsanto's Mind was Changed", *The Guardian*, 9 October 1999.

24. Rockefeller Foundation, *"Terminator" Seed Technology Dropped*, Press Release, New York, 4 October 1999.

CHAPTER 2
The Fox Guards the Hen-House

Science Bending the Knee to Politics

A s GM seeds were being commercially introduced into Argentine and North American farming, an event of enormous significance for the future of the GMO project occurred in faraway Scotland. There, in Aberdeen, in a state-supported laboratory, the Rowett Research Institute, an experienced scientist was making studies in a carefully controlled manner. His mandate was to conduct long-term research on the possible effects of a GMO food diet on animals.

The scientist, Dr. Arpad Pusztai, was no novice in genetic research. He had worked in the specialized field of biotechnology for more than 35 years, published a wealth of recognized scientific papers, and was considered the world's leading expert on lectins and the genetic modification of plants.

In 1995, just prior to Monsanto's mass commercial sales of GM soybean seeds to American and Argentine farmers, Scotland's Office of Agriculture, Environment and Fisheries contracted the Rowett Research Institute to undertake a three-year comprehensive study under the direction of Dr. Pusztai. With a budget of $1.5 million, it was no small undertaking.[1]

The Scottish Agriculture Office wanted Rowett to establish guidelines for a scientific testing methodology to be used by Government regulatory authorities to conduct future risk assessments of GMO crops. As the spread of GMO crops was in its earliest stages, mostly in test or field trials, it was a logical next step to prepare such sound regulatory controls.

No better person could have been imagined to establish scientific credibility, and a sound methodology than Dr. Pusztai. He and his wife, Dr. Susan Bardocz, also a scientist at Rowett, had jointly published two books on the subject of plant lectins, on top of Pusztai's more than 270 scientific articles on his various research findings. He was regarded by his peers as an impeccable researcher.

More significant, in terms of what followed, the Pusztai research project was the very first independent scientific study on the safety of gene-modified food in the world. That fact was astonishing, given the enormous importance of the introduction of genetically modified organisms into the basic human and animal diet.

The only other study of GM food effects at the time was the one sponsored by Monsanto, wherein conclusions not surprisingly claimed that genetically-engineered food was completely healthy to consume. Pusztai knew that a wholly independent view was essential to any serious scientific evaluation, and necessary to create confidence in such a major new development. He himself was fully certain the study would confirm the safety of GM foods. As he began his careful study, Pusztai believed in the promise of GMO technology.

Pusztai was given the task of testing laboratory rats in several selected groups. One group would be fed a diet of GM potatoes. The potatoes had been modified with a lectin which was supposed to act as a natural insecticide, preventing an aphid insect attack on the potato crops—or so went the genetically engineered potato maker's claim.

A Bomb Falls on the GMO Project
The Scottish government, Rowett and Dr. Pusztai believed they were about to verify a significant breakthrough in plant science

which could be of huge benefit to food production by eliminating need for added pesticides in potato planting. By late 1997, Pusztai was beginning to have doubts.

The rats fed for more than 110 days on a diet of GM potatoes had marked changes to their development. They were significantly smaller in size and body weight than ordinary potato-fed control rats in the same experiment. More alarming, however, was the fact that the GMO rats showed markedly smaller liver and heart sizes, and demonstrated weaker immune systems. The most alarming finding from Pusztai's laboratory tests, however, was the markedly smaller brain size of GMO-fed rats compared with normal potato-fed rats. This later finding so alarmed Pusztai that he chose to leave it out when he was asked to present his findings on a UK Independent Television show in 1998. He said later he feared unleashing panic among the population.

What Dr. Arpad Pusztai did say when he was invited to talk briefly about his results on the popular ITV "World in Action" broadcast in August 1998, was alarming enough. Pusztai told the world, "We are assured that this is absolutely safe. We can eat it all the time. We must eat it all the time. There is no conceivable harm which can come to us." He then went on to issue the following caveat to his millions of viewers. He stated, "But, as a scientist looking at it, actively working in the field, I find that it is very, very unfair to use our fellow citizens as guinea pigs. We have to find guinea pigs in the laboratory."

Pusztai, who had cleared his TV appearance beforehand with the director of Rowett, had been told not to talk in detail about his experiments. What he went on to say, however, detonated the political equivalent of a hydrogen bomb across the world of biotechnology, politics, science and GMO agribusiness.

Pusztai stated simply that, "the effect (of a diet of GM potatoes) was slight growth retardation and an effect on the immune system. One of the genetically modified potatoes, after 110 days, made the rats less responsive to immune effects." Pusztai added a personal note: "If I had the choice, I would certainly not eat it until I see at least comparable scientific evidence which we are producing for our genetically modified potatoes."[2]

Suddenly, the world was debating the sensational Pusztai comments. Damage to organs and immune systems was bad enough. But the leading UK gene scientist had also said he himself would not eat GMO food if he had a choice.

The initial response from Pusztai's boss, Prof. Philip James, was warm congratulations for the way Pusztai presented his work that day. On James's decision, the Institute even issued a press release based on Pusztai's findings, stressing that "a range of carefully controlled studies underlie the basis of Dr. Pusztai's concerns."[3]

That token support was to break radically. Within 48 hours, the 68-year-old researcher was told his contract would not be renewed. He was effectively fired, along with his wife, who had herself been a respected Rowett researcher for more than 13 years. Moreover, under the threat of losing his pension, Pusztai was told not to ever speak to the press about his research. His papers were seized and placed under lock. He was forbidden to talk to members of his research team under threat of legal action. The team was dispersed. His phone calls and e-mails were diverted.

That was to be only the beginning of a defamation campaign worthy of Third Reich Germany or Stalinist Russia, both of which Pusztai had survived as a young man growing up in Hungary.

Pusztai's colleagues began to defame his scientific repute. Rowett, after several different press releases, each contradicting the previous, settled on the story that Pusztai had simply "confused" the samples from the GMO rats with those from ordinary rats who had been fed a sample of potato known to be poisonous. Such a basic error for a scientist of Pusztai's seniority and proven competence was unheard of. The Press claimed it was one of the worst errors ever admitted by a major scientific institution.

However, it was simply not true, as a later audit of Pusztai's work proved.

Rowett, according to exhaustive research by UK journalist Andrew Rowell, later shifted its story, finding a flimsy fallback in the claim that Pusztai had not carried out the long-term tests needed to prove the results.

But the clumsy efforts of Prof. James and Rowett Institute to justify the firing and defaming of Pusztai were soon forgotten, as other scientists and government ministers jumped into the frenzy to discredit Pusztai. In defiance of these attacks, by February 1999, some 30 leading scientists from 13 countries had signed an open letter supporting Pusztai. The letter was published in the London *Guardian*, triggering a whole new round of controversy over the safety of GMO crops and the Pusztai findings.

Blair, Clinton and "Political" Science

Within days of the *Guardian* piece, no less august an institution than the British Royal Society entered the fray. It announced its decision to review the evidence of Pusztai. In June 1999, the Society issued a public statement claiming that Pusztai's research had been "flawed in many aspects of design, execution and analysis and that no conclusions should be drawn from it."[4]

Coming from the 300-year-old renowned institution, that statement was a heavy blow to Pusztai's credibility. But the Royal Society's remarks on Pusztai's work were also recognizable as a political smear, and one which risked tarnishing the credibility of the Royal Society itself. It was later revealed by a peer review that the latter had drawn its conclusions from incomplete data. Furthermore, it refused to release the names of its reviewers, leading some critics to attack the Society's methods as reminiscent of the medieval Star Chamber.[5]

Research by Andrew Rowell revealed that the Royal Society's statements and the British House of Commons Science and Technology Select Committee's similar condemnation issued on the same day, May 18, were the result of concerted pressure on those two bodies by the Blair Government.

The Blair Government had indeed set up a secret Biotechnology Presentation Group to launch a propaganda campaign to counter the anti-GMO media, at that point a dominant voice in the UK. The Pusztai debate threatened the very future of a hugely profitable GMO agribusiness for UK companies.

Three days after the coordinated attacks on Pusztai's scientific integrity from the Royal Society and the Select Committee, Blair's so-called "Cabinet Enforcer," Dr. Jack Cunningham, stood in the House of Commons to declare, "The Royal Society this week convincingly dismissed as wholly misleading the results of some recent research into potatoes, and the misinterpretation of it—There's no evidence to suggest that any GM foods on sale in this country are harmful." Making his message on behalf of the Blair Cabinet unmistakable, he added, "Biotechnology is an important and exciting area of scientific advance that offers enormous opportunities for improving our quality of life."[6]

Public documents later revealed that the Blair Cabinet was itself split over the GMO safety issue and that some members advised further study of potential GMO health risks. They were silenced, and Cunningham was placed in charge of the Government's common line on GMO crops, the Biotechnology Presentation Group.

What could possibly explain such a dramatic turnaround on the part of James and the Rowett Institute? As it turned out, the answer was political pressure.

It took five years and several heart attacks, before the near-ruined Pusztai was able to piece together the details of what had taken place in those 48 hours following his first TV appearance in 1998. His findings revealed the dark truth about of the politics of GMO crops.

Several former colleagues at Rowett, who had retired and were thus protected from possibly losing their jobs, privately confirmed to Pusztai that Rowett's director, Prof. Philip James, had received two direct phone calls from Prime Minister Tony Blair. Blair had made clear in no uncertain terms that Pusztai had to be silenced.

James, fearing the loss of state funding and worse, proceeded to destroy his former colleague. But the chain did not stop at Tony Blair. Pusztai also learned that Blair had initially received an alarmed phone call from the President of the United States, Bill Clinton.

Blair was convinced by his close friend and political adviser, Clinton, that GMO agribusiness was the wave of the future, a huge—and growing—multibillion dollar industry in which Blair could offer British pharmaceutical and biotech giants a leading role. What is more, Blair had made the promotion of GMO a cornerstone of his successful 1997 election campaign to "Re-brand Britain." And it was well-known in the UK that Clinton had initially won Blair over to the promise of GM plants as the pathway towards a new agro-industrial revolution.[7]

The Clinton Administration was in the midst of spending billions to promote GMO crops as the technology of a future biotech revolution. A Clinton White House senior staff member stated at the time that their goal was to make the 1990's, "the decade of the successful commercialization of agricultural biotechnology products." By the late 1990's, the stocks of biotech GM companies were soaring on the Wall Street stock exchange. Clinton was not about to have some scientist in Scotland sabotage his project, nor clearly was Clinton's good friend Blair.

The final piece of the puzzle fell into place for Pusztai, thanks to further information from former colleague, Professor Robert Orskov, a leading nutrition scientist with a 33 year career at Rowett. Orskov, who had in the meantime left the institute, told Pusztai that senior Rowett colleagues had informed him that the initial phone call behind his dismissal came from Monsanto.[8]

Monsanto had spoken with Clinton, who in turn had directly spoken to Blair about the "Pusztai problem." Blair then spoke to Rowett's director, Philip James. Twenty four hours later, Dr. Arpad Pusztai was out on the street, banned from speaking about his research and talking to his former colleagues.

Orskov's information was a bombshell. If it was true, it meant that a private corporation, through a simple phone call, had been able to mobilize the President of the United States and the Prime Minister of Great Britain on behalf of its private interests. A simple phone call by Monsanto could destroy the credibility of one of the world's leading independent scientists. This carried somber implications for the future of academic freedom and independent science. But it also had enormous implications for the proliferation of GMO crops worldwide.[9]

A Not-so-ethical Royal Society Joins the Attack

With his scientific reputation already severely damaged, Pusztai finally managed, in October 1999, to secure the publication of his and his colleague's research in the respected British scientific journal, *The Lancet*. The magazine was highly respected for its scientific independence and integrity, and before publication, the article was submitted to a six-person scientific review panel, passing with 4 votes in favor.

The Lancet editor, Dr. Richard Horton, later said he had received a "threatening" phone call from a senior person at the Royal Society, who told him that his job might be at risk should he decide to publish the Pusztai study. Prof. Peter Lachmannn, former Vice President of the Society, later admitted to phoning Horton about the Pusztai paper, though he denied having threatened him.

Investigative journalists from the *Guardian* newspaper discovered that the Royal Society had set up a special "rebuttal unit" to push a pro-GM line and discredit opposing scientists and organizations. The unit was headed by Dr. Rebecca Bowden, a former Blair environment ministry official who was openly pro-GMO.[10]

The paper discovered that Lachmann, who publicly called for scientific "independence" in his attack on Pusztai, was himself hardly an impartial judge of the GMO issue. Lachmann was a scientific consultant for a private biotech company, Geron Biomed, doing animal cloning similar to Dolly the Sheep, and was a non-executive director of the agri-biotech firm, Adprotech. He was also a member of the scientific advisory board of the GMO pharmaceutical giant, SmithKleinBeecham. Lachmann was many things, but impartial in the issue of GMO science he definitely was not.

Lord Sainsbury was the leading financial contributor to Tony Blair's "New Labour" party in the 1997 elections. For his largesse, Sainsbury had been given a Cabinet post as Blair's Science Minister. His science credentials were minimal but he was a major shareholder in two GMO biotech companies, Diatech and Innotech, and was aggressively pro-GMO.

To cement the ties further between the Blair government and leading biotech companies, the PR firm director who successfully ran Blair's 1997 and 2001 election campaigns, Good Relations David Hill, also ran the PR for Monsanto in the UK.

Shedding more doubt on the self-proclaimed scientific neutrality of the Royal Society, was the fact that despite its public pronouncements on Pusztai's "flawed" research, the Society never went on to conduct a "non-flawed" version of the important study. This suggested that their interests lay perhaps in something else than scientific rectitude.

Following the publication of Pusztai's article, *The Lancet* was severely attacked by the Royal Society and the biotechnology industry, whose pressure eventually forced Pusztai's co-author, Prof. Stanley Ewen, to leave his position at the University of Aberdeen.[11]

Science in the Corporate Interest...

The Pusztai case, as devastating as it threatened to be to the entire GMO project, was one among several cases of suppression of independent research or of direct manipulation of research data proving the potentially negative effects of GMO foods on human or animal health. In fact, this practice proved to be the rule.

In 2000, the Blair government ordered a three-year study to be carried out by a private firm, Grainseed, designed to demonstrate which GMO seeds might safely be included on the National List of Seeds, the standard list of seeds farmers may buy.

Internal documents from the UK Ministry of Agriculture were later obtained by the London *Observer* newspaper, and revealed that some strange science was at work in the tests. At least one researcher at the Grainseed firm manipulated scientific data to "make certain seeds in the trials appear to perform better than they really did." Far from causing the Ministry of Agriculture to suspend the tests and fire the employee, the Ministry went on to propose that a variety of GMO corn be certified.[12]

In another example of British state intrusion in academic freedom and scientific integrity, Dr Mae-Wan Ho, senior academic scientist at the Open University and later Director of the Institute of Science in Society, was pressured by her university into taking early retirement. Mae-Wan Ho had been a Fellow of the National

Genetics Foundation in the US, had testified before the UN and World Bank on issues of bioscience, had published widely on genetics, and was a recognized expert on GMO science.

Her "mistake" was that she was too outspoken against the dangers of GMO foods. In 2003, she served on an international Independent Science Panel on GM plants, where she spoke out against the slipshod scientific claims being made about GMO safety.

She warned that genetic modification was entirely unlike normal plant or animal breeding. She stated, "Contrary to what you are told by the pro-GM scientists, the process is not at all precise. It is uncontrollable and unreliable, and typically ends up damaging and scrambling the host genome, with entirely unpredictable consequences." That was more than enough for the GMO lobby to pressure her into "retirement."[13]

To protect the so-called integrity of state-funded research into the safety of GMO foods and crops, the Blair government put together a new code of conduct. Under the Government's Biotechnology and Biological Science Research Council code (BBSRC), any employee of a state-funded research institute who dared to speak out on his findings into GMO plants, could face dismissal, be sued for breach of contract or face a court injunction.

Many institutes doing similar research into GMO foods, such as the John Innes Centre's Sainsbury Laboratory, the UK's leading biotechnology institute, had received major financial backing from GMO biotechnology giants such as Zeneca and Lord Sainsbury personally. As Science Minister, Lord Sainsbury saw to it that the BBSRC got a major increase in government funding in order to carry out its biotech police work of suppressing scientific dissent.

The board of the BBSRC was made up of representatives of large multinationals with a vested interest in the research results, while public interest groups such as the Country Landowners' Association were kept out.[14]

In March 2003, a rare case of dissent took place in the Blair government lobby against allowing the free introduction of virtually untested GMO products into the UK diet. Dr. Brian John submitted a memorandum to the British journal, *GM Science Review*, entitled,

"On the Corruption of GM Science." John stated, "There is no balance in the GM research field or in the peer-review process or in the publication process. For this we have to thank corporate ownership of science, or at least this branch of it. ... Scientific integrity is one loser, and the public interest is another."[15]

Dr. John went on to critique sharply the Royal Society in the area of GMO science, in which "inconvenient research simply never sees the light of day." He added, "The prevention of academic fraud is one matter; the suppression of uncomfortable research results is quite another." John further pointed out that the International Life Sciences Institute Bibliography on GMO safety investigations was overwhelmingly biased towards pro-GMO papers, either from Government sources or directly from the biotech industry themselves. "Very few of them involve genuine GM feeding trials involving animals, and none of them so far as I can see, involves feeding trials on humans."[16]

Pusztai's research at the Rowett Institute was one of the first and last in the UK to involve live animal research. The Blair government was determined not to repeat that mistake. In June 2003, amid the furor in the British House of Commons over the decision to back George W. Bush's war in Iraq, Tony Blair sacked his Environment Minister, Michael Meacher. Meacher, later openly opposing UK involvement in Iraq, was in charge of his Ministry's three-year study of GMO plants and their effects on the environment. Openly critical of the prevailing research on GMO crops, Meacher had called on the Blair government to make far more thorough tests before releasing GMO crops for general use. As Mr. Meacher was becoming an embarrassment to the genetic revolution, the response was the French Revolution's—"Off with his head."

As determined the Blair government was in its support of the GMO revolution, its efforts paled in comparison to those of its closest ally across the Atlantic. The United States, the cradle of the GMO revolution in world agriculture, was way ahead of the game in terms of controlling the agenda and the debate. The US GMO campaign of the 1980's and 1990's however had roots in policies going decades back. Its first public traces were found during the

Vietnam War era of the late 1960's and into the second Nixon Presidency. Henry Kissinger, a Rockefeller protégé, was to play a decisive role in that early period. He had introduced the idea of using "food as a weapon" into United States foreign policy. The "food weapon" was subsequently expanded into a far-reaching US policy doctrine.

Notes

1. Author's interview with Dr. Pusztai, 23 June 2007.

2. *Ibid.*

3. The exact words were "the rats had slightly stunted growth when tested after 110 days of feeding and the response of their lymphocytes to mitogenic stimuli was about half that of controls." A second press release from the Chairman of the Institute's governing body, published on 10 August 1998, the same day as the ITV's *World in Action* TV interview with Pusztai, asked for an assurance from the European Commission "that any GMOs be adequately tested for any effects that might be triggered by their consumption in animals or humans". In addition, "The testing of modified products with implanted genes needs to be thoroughly carried out in the gut of animals if unknown disasters are to be avoided," cited in Alan Ryan et al., *Genetically Modified Crops: the Ethical and Social Issues*, Nuffield Council on Bioethics, pp. 140-141.

4. The Royal Society, *Review of Data on Possible Toxicity of GM Potatoes*, June 1999, Ref: 11/99, p. 1, in http://www.royalsoc.ac.uk.

5. The Royal Society itself had extensive ties to the corporate sponsorship of industrial biotech firms such as Aventis Foundation, BP plc, Wellcome Trust, Astra-Zeneca plc, Esso UK plc, the Gatsby Charitable Foundation, Andrew W. Mellon Foundation, cited in Martin J. Walker, *Brave New World of Zero Risk: Covert Strategy in British Science Policy*, London, Slingshot Publications, 2005, pp. 173-193.

6. Jack Cunningham, Minister for the Cabinet Office, *Statement to House of Commons*, 21 May 1999, in http://www.publications.parliament.uk/pa/cm199899/cmhansrd/vo990521/debtext/90521-07.htm.

7. Tony Blair, press comment, *Remarks Prior to Discussions With Prime Minister Tony Blair of the United Kingdom and an Exchange With Reporters in Okinawa—Transcript*, Weekly Compilation of Presidential Documents, 31 July 2000, in http://www.gpoaccess.gov/wcomp/. Blair's comments during his meeting with Clinton then were: "…this whole science of biotechnology is going—I mean, I'm not an expert on it, but people tell me whose opinions I respect that this whole science of biotechnology is perhaps going to be, for the first half of the 21st century what information technology was to the last half of the 20th century. And therefore, it's particularly important, especially for a country like Britain that is a leader in this science of biotechnology…"

8. Robert Orskov, quoted in Andrew Rowell, "The Sinister Sacking of the World's Leading GM Expert—and the Trail that Leads to Tony Blair and the White House", *The Daily Mail*, 7 July 2003.

9. Andrew Rowell, *Don't Worry, it's Safe to Eat: The True Story of GM food*, BSE and Foot and Mouth, London, 2003, and Rowell, "The Sinister Sacking..." op. cit. Arpad Pusztai, Letter from Arpad Pusztai to the Royal Society dated 12/05/1999, provides a personal account of the scientific events, in http://www.freenet-pages.co.uk/hp/A.Pusztai/RoyalSoc/Pusztai...htm. The official Rowett Institute version of the Pusztai events is on http://www.rowett.ac.uk/gmoarchive. The same site reproduces the entire Pusztai 1998 analysis of the GMO potato experiments with rats, *SOAEFD Flexible Fund Project RO 818: Report of Project Coordinator on Data Produced at the Rowett Research Institute (RRI)*, 22 October 1998. Arpad Pusztai, "Why I Cannot Remain Silent", GM-FREE magazine, August/September 1999. Subsequent to Pusztai's firing, he sent the research protocols to 24 independent scientists in different countries. They rejected the conclusions of the Review Committee and found that his research was of good quality and defended his conclusions. Their report was ignored in media and government circles.

10. Laurie Flynn and Michael Sean, "Pro-GM Scientist 'Threatened Editor'", *The Guardian*, 1 November 1999.

11. Stanley Ewen and Arpad Pusztai, "Effect of Diets Containing Genetically Modified Potatoes Expressing Galanthus Nivalis Lectin on Rat Small Intestine", *The Lancet*, 16 October 1999. A detailed scientific defense of Pusztai's work was given by former colleague, T. C. Bog-Hansen, who became senior associate professor at the University of Copenhagen. See http://plab.ku.dk/tcbh/Pusztai. Geoffrey Lean, "Expert on GM Danger Vindicated", *The Independent*, 3 October 1999. For a thorough account of the witchhunt against Pusztai: George Monbiot, "Silent Science", in *Captive State: The Corporate Takeover of Britain*, Pan Books, London, 2000.

12. Anthony Barnett, "Revealed: GM Firm Faked Test Figures," *The Observer*, 16 April 2000.

13. Anastasia Stephens, "Puncturing the GM Myths", *The Evening Standard*, 8 April 2004. Despite the pressure, Dr. Mae-Wan Ho continued to be one of the few scientists to speak out on the dangers of GMO plants.

14. Norfolk Genetic Information Network, *Scientists Gagged on GM Foods by Public Funding Body with Big Links to Industry*, press release, 1999, http://www.ngin.tripod.com/scigag.htm.

15. Dr. Brian John, "On the Corruption of GM Science", Submission to the *GM Science Review*, 20 March 2003. The UK Government closed the journal, curiously enough, in 2004. It had been founded in 2002 to deepen debate on the issue of GMO plants.

16. *Ibid.*

PART II
The Rockefeller Plan

CHAPTER 3
"Tricky" Dick Nixon and Trickier Rockefellers

America's Vietnam Paradigm Shift

W hen Richard Nixon stepped into the White House as President in January 1969, the United States of America was in a deep crisis. A very select few saw the crisis as a long-awaited opportunity. Most Americans, however, did not.

For the next six years, Nixon was to preside over the first major military defeat ever suffered by the United States, the loss of the war in Vietnam. Hundreds of thousands of American students were marching on Washington in protest against a war which seemed utterly senseless. Morale among young Army draftee soldiers in Vietnam had collapsed to an all-time low, with drug addiction rampant among GI's, and enraged rebellious soldiers "fragging" or killing their company commanders in the field. America's youth were being brought back home in body bags by the thousands. In those days the Pentagon still allowed the press to photograph the returning dead.

The US economy was in severe shock. It was the first time its post-war superiority was being eclipsed by newer and more efficient industries in Western Europe and Japan. By 1969, as Nixon took office, the US dollar itself had entered a terminal crisis as foreign central banks demanded gold instead of paper dollars for their growing trade surpluses with the United States. The post-war profit rate of American corporations, which had peaked in 1965, was now in steady decline.

American corporations found that they could make higher profits by going abroad and buying foreign companies. It was the beginning of significant American corporate multi-nationalism, the precursor to the later phenomenon of globalization. US jobs were disappearing in traditional domestic industries, and the Rust Belt was spreading across once-thriving steel producing regions. The post-war pillar of American industrial superiority was vanishing, and fast.

American industry was rusting as its factories, most of which were built before and during the war, had become obsolescent compared with the modern new post-war industry in Western Europe and Japan. Corporate America faced severe recession and its banks were hard pressed to find profitable areas for lending.

From 1960 to 1974, debt began to grow at an explosive rate in every corner of the US economy. By 1974, corporate debt, home mortgage debt, consumer debt and local government debt had risen by a combined 300%. During the same 15-year period, the debt of the US Government had risen by an even more impressive 1,000%. By the early 1970's, the United States was by every traditional measure in a deep economic crisis. Little wonder there was growing skepticism abroad that the US dollar would continue to hold its value against gold.

Within a quarter century after the 1944 creation of the Bretton Woods monetary system, the establishment's version of an American Century dominating world affairs was rapidly running up against fundamental problems, problems which led to bold new searches among the US establishment and its wealthiest families, for new areas of profit.

Food or, as it was about to be renamed, US agribusiness, was to become a vital pillar of a new American economic domination by the 1960's, along with a far more expensive petroleum. It was a paradigm shift.[1]

The Vietnam War and its divisive social impact were to last until the humiliating resignation of Nixon in August 1974, a losing victim of power struggles within the US establishment.

No figure had played a more decisive role in those power plays than former New York Governor, Nelson Rockefeller, a man who desperately wanted to be President if he could. To reach that goal in the midst of Nixon's crisis was in fact the main aim of Nelson Rockefeller. Rockefeller, together with his brothers David, Laurance, John, and Winthrop, ran the family's foundation along with numerous other tax-exempt entities such as the Rockefeller Brothers Fund.

At the beginning of the crisis-torn 1970's, certain influential persons within the American establishment had clearly decided a drastic shift in direction of US global policy was in order.

The most influential persons were David and Nelson Rockefeller, and the group of influential political and business figures around the Rockefeller family. The family's power center was the exclusive organization created in the aftermath of World War I, the New York Council on Foreign Relations.

In the 1960's the Rockefellers were at the power center of the US establishment. The family and its various foundations dominated think-tanks, academia, government and private business in the 1960's in a manner no other single family in United States history had managed to then. Secretary of State Henry Kissinger had been their hand-picked protégé, recruited from Harvard in the late 1950's to work for a new Rockefeller Foundation project.[2]

David Rockefeller's "Crisis of Democracy"
One response by the US establishment inner-circles to the late-1960's crisis in the American hegemony, was a decision to create a new division of the global economic spoils, for the first time inviting Japan into the "rich-mens' club".

In 1973, following a meeting of some 300 influential, hand-picked friends of the Rockefeller brothers from Europe, North America and Japan, David Rockefeller expanded the influence of his establishment friends and founded a powerful new global policy circle, the Trilateral Commission. The "triangle" included North America, Europe and now, Japan.

Among the 1973 founding members of David Rockefeller's Trilateral Commission were Zbigniew Brzezinski, and a Georgia Governor and peanut farmer, James Earl "Jimmy" Carter, along with George H.W. Bush, Paul Volcker, later named by President Jimmy Carter as Federal Reserve chairman, and Alan Greenspan, then a Wall Street investment banker. It was no small-time operation.

The idea of a new, top organization similar to the US Council on Foreign Relations, incorporating not only Western European policy elites, but also Japan for the first time, grew from talks between David Rockefeller and his Maine neighbor, Zbigniew Brzezinski. Brzezinski was then Professor at Columbia University's Russian Studies Center, a recipient of generous Rockefeller Foundation funding.

Brzezinski had just written a book where he proposed the idea of consolidation of American corporate and banking influence worldwide via a series of regular closed-door policy meetings between the select business elites of Europe, North America and Japan.

His personal views were not exactly the stuff of traditional American democracy and liberty. In this little-known book, *Between Two Ages: America's Role in the Technetronic Era*, published in 1970, Brzezinski referred to the significant policy voices in the United States as, "the ruling elite," stating bluntly that, "Society would be dominated by an elite ... [which] would not hesitate to achieve its political ends by using the latest modern techniques for influencing public behavior and keeping society under close surveillance and control."

Brzezinski was chosen by David Rockefeller to be the first Executive Director for Rockefeller's Trilateral Commission.

The Trilateral Commission, a private elite organization, laid the basis of a new global strategy for a network of interlinked international elites, many of them business partners of the Rockefellers, whose combined financial, economic and political weight was unparalleled. Its ambition was to create what Trilateral member George H.W. Bush later called a "new world order," constructed on the designs of Rockefeller and kindred wealthy interests. The Trilateral group laid the foundation of what by the 1990's came to be called "globalization."

One of the first policy papers issued by David Rockefeller's Trilateral Commission group was drafted by Harvard Professor Samuel Huntington, the person who was to draft a controversial "Clash of Civilizations" thesis in the mid-1990's, which laid the basis for the later Bush Administration War on Terror.

The 1975 Huntington report was titled: "The Crisis of Democracy."[3]

For Huntington and David Rockefeller's establishment associates at the Trilateral Commission, the "crisis," however, was the fact that hundreds of thousands of ordinary American citizens had begun to protest their government's policies. America, or at least its power elite, was threatened, Huntington declared by an "excess of democracy." The unruly "natives" were clearly getting too "restless" for the elite circles of the establishment around Huntington and David Rockefeller.

Huntington went on to warn, "The effective operation of a democratic political system usually requires some measure of apathy and non-involvement on the part of some individuals and groups." He also insisted that, "... secrecy and deception ... are ... inescapable attributes of ... government."[4]

The unreliable nature of democratic governments, subject to the pressures of an unpredictable popular mood, only demonstrated for these circles around Huntington and David Rockefeller's Trilateral Commission the wisdom of, among other things, privatizing public enterprise and deregulating industry. The movement to deregulate and privatize government services actually began under President Jimmy Carter, a hand-picked David Rockefeller candidate, and a Trilateral Commission founding member.

This was hardly the song of "America, the Beautiful." The document was an alarm call from the US power establishment and its wealthy patrons. Drastic situations required drastic measures.

Kissinger and Food Politics

Henry Kissinger was moving to take complete control over the US foreign policy apparatus by 1973.

And as both Secretary of State and the President's National Security Adviser, Kissinger was to make food a centerpiece of his diplomacy along with oil geopolitics.[5]

Food had played a strategic, albeit less central role in post-war US foreign policy with the onset of the Cold War. It was masked under the rhetoric of programs with positive sounding names, such as Food for Peace (P.L. 480). Often Washington claimed its food exports subsidies were tied to domestic pressure from its farmers. This was far from the real reason, but it served to cover the true situation, that American agriculture was in the process of being transformed from family-run small farms to global agribusiness concerns.

Domination of global agriculture trade was to be one of the central pillars of post-war Washington policy, along with domination of world oil markets and non-communist world defense sales. Henry Kissinger reportedly declared to one journalist at the time, "If you control oil, you control nations. If you control food, you control people."

By the early 1970's, Washington, or more accurately, very powerful private circles, including the Rockefeller family, were about to try to control both, in a process whose daunting scope was perhaps its best deception.

Initially, the agriculture weapon was used by Washington more as a club to hit other countries. Starting in the 1970's, there was a major shift in food policy. This redirection was a precursor to the agro-chemical cartels of the 1970's.

The defining event for the emergence of a new US food policy was a world food crisis in 1973, which took place at the same time that Henry Kissinger's "shuttle diplomacy" triggered the OPEC 400% increase in world oil prices. The combination of a drastic energy price shock and a global food shortage for grain staples, was the breeding ground for a significant new Washington policy turn. The turn was wrapped in "national security" secrecy.

In 1974 the United Nations held a major UN World Food Conference in Rome. The Rome conference discussed two main themes, largely on the initiative of the United States. The first was supposedly alarming population growth in the context of world food shortages, a one-sided formulation of the problem. The second theme was how to deal with sudden changes in world food supply and rises in prices. Prices for oil and grains were both rising on international markets at annual rates of 300 to 400% at that time.

A convenient if unintended consequence of the food crisis, was a strategic increase in the geopolitical power of the world's largest food surplus producer, the United States, over the world food supply and, hence, global food prices. It was during this time that a new alliance grew up between private US-based grain trading companies and the US Government. That alliance laid the ground for the later gene revolution.

The "Great Grain Robbery"

As Secretary of State, Henry Kissinger had made an internal power play to control US agriculture policy, traditionally the domain of the US Department of Agriculture. Kissinger did this through his role in negotiating huge US grain sales to the Soviet Union in exchange for Russian oil, in the months before the Rome Food Conference.

The Soviets agreed to buy an unprecedented 30 million tons of grain from the United States under the Kissinger deal. The amounts were so huge that Washington turned to the private grain traders like Cargill, not to its usual Government reserves, to sell Russia the needed grain. That was part of the Kissinger plan. As an aide to Kissinger explained at the time, "Agriculture policy is too important to be left in the hands of the Agriculture Department."

The Soviet grain sale was so large that it depleted world reserves and allowed the trading companies to raise wheat and rice prices by 70% and more in a matter of months. Wheat went from $65 a ton to $110 a ton. Soybean prices doubled. At the same time, severe drought had cut the grain harvests in India, China, Indonesia,

Bangladesh, Australia and other countries. The world was desperate for imported grain and Washington was preparing to take advantage of that desperation to radically change world food markets and food trade.

The deal was called the "great grain robbery" in reference to the overly friendly terms of sale to Moscow and the low price that year paid to US farmers for the same grain. Kissinger had negotiated the Soviet sale with the enticement of generous US Export-Import Bank credits and other subsidies.[6]

The big winners were the US-based grain traders such as Cargill, Archer Daniels Midland, Bunge and Continental Grain, who emerged as true global agribusiness giants. Kissinger's new food diplomacy created a global agriculture market for the first time. This potential for power and control over whole areas of the planet was not lost on the US establishment, least of all, not on Kissinger.

In 1974 the world was reeling under the shock of the 400% increase in world oil prices, a shock Kissinger had more than a little to do with from behind the scenes.[7]

During this period, as world oil prices were going through the roof, there was a catastrophic world harvest failure. The Soviet grain harvest had been devastated through crop failure and other problems. The United States was the world's only major surplus supplier of wheat and other agriculture commodities. It marked a major shift for Washington agriculture export policy.

Kissinger was both Secretary of State and National Security Adviser to the President in early 1974. The Secretary of Agriculture was Earl Lauer Butz, a friend of agribusiness, an avid promoter of population control, a racist whose remarks about blacks cost him his job, and who later was convicted for tax evasion.

Time magazine on November 11, 1974, concluded a special report on the world food crisis, explaining why they were in favor of triage, the wartime practice of deciding which war-wounded can survive and which left to die:

In the West, there is increasing talk of triage.... If the US decides that the grant would simply go down the drain as a mere palliative because the recipient country was doing little to improve its food distribution or start a population control program, no help would be sent. This may be a brutal policy, but it is perhaps the only kind that can have any long-range impact. A triage approach could also demand political concessions....Washington may feel no obligation to help countries that consistently and strongly oppose it. As Earl Butz told Time, "Food is a weapon. It is now one of the principal tools in our negotiating kit."[8]

Providing food, however, was not to be the real weapon. The denial of food—famine—was.

"When in Rome..."

During the Cold War, Washington consistently opposed the creation of internationally held grain reserves. The virtual depletion of world food stocks prompted the 1974 UN World Food Conference meeting in Rome. In 1972, when the world suffered an exceptionally poor harvest, there were 209 million metric tons of grain, some 66 days' worth, in world reserve. In 1974 there were record grain crops worldwide, yet the grain reserve was reduced to 25 million metric tons, or 37 days. In 1975 there was estimated to be a 27-day reserve after exceptionally large grain harvests.[9]

The problem was that the grain was there, but it was in the hands of a handful of giant grain trading companies, all of them American. This was the element that Kissinger had in mind when he spoke of food as a weapon.

George McGovern, chairman of the Senate Committee on Human Nutrition, stated at the time, "Private traders are in business to turn investments into profit as rapidly as possible ... In reality a reserve in private hands is no reserve at all. It is indeed precisely the same market mechanism which has produced the situation we face today."[10]

McGovern was not appreciated by the US establishment for such comments. His bid against Nixon for the Presidency in 1972 was doomed to be a disastrous defeat for traditional elements in the

Democratic Party. The trading giants were deliberately manipulating available grain supply to hike prices. Because the US government required no accurate grain reporting, only the grain giants like Cargill and Continental Grain knew what they had.

James McHale, Pennsylvania's Secretary of Agriculture, had gone to Rome in 1974 to plead for a sensible international food policy. He pointed out that 95 per cent of all grain reserves in the world at the time were under the control of six multinational agribusiness corporations—Cargill Grain Company, Continental Grain Company, Cook Industries Inc., Dreyfus, Bunge Company and Archer-Daniel Midland. All of them were American-based companies.[11]

This connection between Washington and the grain giants was the heart of Kissinger's food weapon. Jean Pierre Laviec of the International Union of Food Workers said in a statement released at the Rome Food Conference, referring to the Big Six, "They decide the quantities of vital inputs to be produced, the quantities of agricultural products to be bought, where plants will be built and investments made. The growth rate of agribusiness has risen during the last ten years and ... has been directly proportional to the increase of hunger and scarcity."[12]

What was to come in the following ten years and more would far surpass what Laviec warned of in 1974. The United States was about to reorganize the global food market along private corporate lines, laying the background for the later "Gene Revolution" of the 1990's.

No one group played a more decisive role in that reshaping of global agriculture during the next two decades than the Rockefeller interests and the Rockefeller Foundation.

Nixon's Agriculture Export Strategy

The emergence of a US-dominated global market in grain and agriculture commodities was part of a long-term US strategy which began in the early 1970's under Richard Nixon. In August 1971, Nixon had taken the dollar off the gold exchange standard of the 1944 Bretton Woods monetary system. He let it devalue in a freefall, or float, as it was called. This was part of a strategy which included making US grain exports strategically competitive in Europe and around the world.

Free trade was the war cry of the Nixon Administration. Cargill, Continental Grain, Archer Daniels Midland were its new warriors.

In 1972, William Pearce became Nixon's Special Deputy Representative for Trade Negotiations, with rank of Ambassador. He had been one of the chief policy members of the President's Commission on International Trade and Investment Policy, a special trade group chaired by the former President of IBM, Albert Williams. At that time Pearce was Cargill's Vice President for Public Affairs.

Not surprisingly, Pearce saw to it that the final Williams Commission report to the President recommended that the US should pressure other countries to eliminate agriculture trade barriers which blocked imports of US agriculture products, and it argued against policies that supported what Pearce preferred to call "inefficient farmers." Pearce ensured that the Williams report focused on how to expand US food exports.

Some years later, Cargill Vice Chairman, Walter B. Saunders, told a National Grain and Feed Association convention in New Orleans, "The fundamental problem with farm policy goes back nearly 50 years to the belief that the best way to protect farm income is to link it to price.... Income must become less dependent on unit prices and more dependent of production efficiencies, diversification of income sources, better marketing and greater volume."[13] In clear words, the family farmer had to get out of the way and let the new giant agribusiness conglomerates dominate.

This policy shift, all in the name of the American virtue of "efficiency," was to have fateful consequences over the following three decades.

Cargill's Pearce argued that American agriculture had unique advantages of scale and efficiency, technology and capital, which made it the natural contender for world export leader. Countries trying to defend their own farmers such as the European Economic Community, by his argument, were defending "inefficiency." Washington was out to dismantle the European Common Agriculture Policy, the backbone of France's post-war political stability.

The Williams-Pearce report used the global security umbrella argument, pointing out, "many of the economic problems we face today grow out of the overseas responsibilities the US has assumed as the major power of the non-communist world." It forgot to mention the deliberate background to that United States global "policeman" role. It was a thinly veiled argument to justify US pressure on its trading partners to open their markets to Cargill and other US agribusiness giants, to "help repay" the US for its Cold War role.

Pearce's strategy became a central part of the 1972 Nixon New Economic Policy. Two years later, Cargill's Pearce was on the President's Committee for Economic Development, where he developed domestic US agriculture policy. There his target was to remove US farming's "excess human resources" (*sic*), to drive hundreds of thousands of family farmers into bankruptcy to make room for vast agribusiness farming. He then went back to Cargill, yet another practitioner of the revolving door system between select private companies and the Government agencies they depend on.

The Pearce strategy, adopted by the Nixon Administration, was a thinly veiled form of food imperialism. Europe, Japan and other industrialized countries should give up their domestic agriculture self-sufficiency support, and open the way for the United States to become the world granary, the most "rational" use of world resources. Anything else was patently "inefficient."

Washington would use the classic British "free trade" argument, in play since the 1846 Repeal of the Corn Laws, where the dominant economic and trade power benefits from forcing removal of trade protection of weaker competitors.

Pearce's, or more accurately, Cargill's strategy was to shape US trade policy for the following three decades, and play a decisive role in the ability of a handful of giant American agri-chemical corporations to take over the world market in seeds and pesticides with their GMO plants.

In order to become the world's most efficient agriculture producer, Pearce argued, traditional American family-based farming must give way to a major revolution in production. The family farm was to become the "factory farm," and agriculture was to become "agribusiness."

The Williams Commission believed that to carry out such "free trade" policies, US agriculture would have to be converted into an efficient export industry, phasing out domestic farm programs designed to protect farm income and move to a "free market" oriented agriculture. This approach was widely supported by corporate agribusiness, big New York banks and investment firms who saw the emerging agribusiness as a potential group of new "hot" stocks for Wall Street. It became the cornerstone of the Nixon Administration's farm policy.

Agribusiness and international trading giants like Cargill and Archer Daniels Midland (ADM), would set the priorities of US agriculture policy. The idea of US food self-sufficiency was replaced by a simple motto: what's good for Cargill and the grain export trading companies was "good for American agriculture." The family farmer got lost somewhere in the shuffle, along with his Senate champion, George McGovern.

By devaluing the dollar in August 1971, and adopting his New Economic Plan (NEP), Nixon took a first step in carrying out the new export policies. As the president of the National Grain and Feed Association described it, "the NEP was very important in giving US agriculture an advantage due to the devaluation of the dollar."[14]

Pearce further argued that Third World countries should give up trying to be food self-sufficient in wheat, rice, and other grains or beef, and focus instead on small fruits, sugar or vegetables. They should import the more efficient US grains and other commodities, naturally shipped by Cargill at prices controlled by Cargill, paying for it by export of the fruit and vegetables. In the bargain they would also lose food self-sufficiency. This was to open a vastly more strategic lever over developing countries over the next three decades, control of their food.

When a poorer or less developed land removed defenses against foreign food imports and opened its markets to mass-produced US products, the results could be predicted, as Pearce and Cargill well knew. According to economist J.W. Smith:

Highly mechanized farms on large acreages can produce units of food cheaper than even the poorest paid farmers of the Third World. When this cheap food is sold, or given, to the Third World, the local farm economy is destroyed. If the poor and unemployed of the Third World were given access to land, access to industrial tools, and protection from cheap imports, they could plant high-protein/high calorie crops and become self-sufficient in food. Reclaiming their land and utilizing the unemployed would cost these societies almost nothing, feed them well, and save far more money than they now pay for the so-called "cheap" imported foods.[15]

But such a sensible alternative was not to be allowed. The Nixon Administration began the process of destroying the domestic food production of developing countries as the opening shot in an undeclared war to create a vast new global market in "efficient" American food exports. Nixon also used the post-war trade regime known as the General Agreement on Tariffs and Trade (GATT) to advance this new global agribusiness export agenda.

In 1972, the Nixon Administration, with Cargill's Pearce in the key post of White House Deputy Trade Representative, and Peter Flanigan as head of Nixon's Council on International Economic Policy, developed the negotiating strategy for the upcoming GATT multilateral trade and tariff negotiations. Their main target in the next phase of their war for domination of world agriculture markets was the Common Agricultural Policy (CAP) of the European Common Market countries, the European Community.[16]

The CAP had been built around protective tariffs when the European Economic Community was first created in the late 1950's, to prevent dumping of US and other agriculture products onto the fragile post-war European market.

Pearce negotiated Congressional passage of the Trade Reform Act of 1974, which directed US negotiators to trade off concessions from the US in the industrial sector, in exchange for concessions to the US in the agricultural sector. This only accelerated the decline of many long-time US industries, like steel, which soon left an unseemly residue in the jobless and abandoned communities of the so-called "rust belt" scattered throughout the northeastern

USA. Steel was called a "sunset" industry, while agribusiness was to become a "sunrise" industry in the parlance of the day.

"Food as a Weapon"

Backed by Cargill and the giant US grain trade conglomerates, Henry Kissinger began an aggressive diplomacy, which he referred to as "Food as a Weapon." The Russian "grain robbery" had been one example of his diplomacy with the food weapon, a "carrot" approach. Another was his use of P.L. 480 in Vietnam during the War.

As popular opposition to the Vietnam War grew, it became difficult for the Administration to get funding from Congress for economic and military aid to South Vietnam. Congress was putting limitations on aid and the White House was looking for ways to avoid this kind of interference. One solution was to divert US aid through multilateral institutions dominated by the US, and another was to use food aid to support US diplomatic and military objectives.

P.L. 480 programs were not subject to annual Congressional appropriations review and Nixon could spend up to $2.5 billion by borrowing from the Department of Agriculture's Commodity Credit Corporation, the same agency used some years later to covertly funnel US military aid to Saddam Hussein. With commercial markets booming and government reserves exhausted the Agriculture Department no longer needed P.L. 480 to dispose of surplus grain and food. The State Department played a major role in determining where the aid went. Kissinger's motto was clearly one of "rewarding friends and punishing enemies".

P.L. 480 became a direct military subsidy for the Indochina war machine. In the beginning of 1974 the food aid to South Vietnam was $207 million. When Congress cut economic aid by 20%, the White House increased the P.L. 480 allocation to $499 million. Kissinger added a special provision so Vietnam and Cambodia could use 100% of counterpart funds for direct military purposes.[17]

When Congress passed an amendment in 1974 requiring that 70% of food aid be given to countries on the UN's list of the Most

Seriously Affected countries, Kissinger tried to get the UN to put South Vietnam on its list, which failed. Ultimately the White House circumvented Congress by just upping the amount of PL 480 aid from $1 billion to $1.6 billion.[18]

Kissinger then aimed his food weapon at Chile.

Like other forms of US aid to Chile, PL 480 was turned "off" when the socialist government of Salvador Allende came into power and began to implement a series of economic reforms. The aid cut-off was done on Kissinger's orders. It was turned back "on" as soon as the US-backed military dictatorship of Augosto Pinochet was in power.

Food played a key part in the Kissinger-orchestrated coup against Allende in 1973. Supported by the State Department and the CIA, right-wing wealthy Chilean landowners sabotaged food production, doubling food imports and exhausting Chile's foreign reserves.[19] This made it very difficult for Chile to import food. The ensuing food shortages created middle class discontent. Allende's request for food credit was denied by the US State Department, even though it should have been the Department of Agriculture's domain. Kissinger had stolen the turf from Agriculture Secretary Earl Butz.

After the 1973 military coup, the US food aid granted to Chile was sold on the domestic market by the Pinochet government. That did nothing to ease the plight of the workers there because of massive inflation and erosion of purchasing power. The military junta was the main beneficiary because the influx of food aid eased balance of payments difficulties and freed up money for the military, at the time the 9th largest importer of US arms.[20]

Back in 1948, as the Cold War was heating up, and Washington was setting up NATO, the man who was the architect of the US policy of "containment" of the Soviet Union, State Department senior planning official George Kennan noted in a Top Secret memorandum to the Secretary of State:

> We have about 50% of the world's wealth but only 6.3% of its population.... In this situation, we cannot fail to be the object of envy and resentment. Our real task in the coming period is to devise a pattern of relationships which will permit us to maintain this position of dis-

parity without positive detriment to our national security. To do so, we will have to dispense with all sentimentality and day-dreaming; and our attention will have to be concentrated everywhere on our immediate national objectives. We need not deceive ourselves that we can afford today the luxury of altruism and world-benefaction.[21]

Such a steel-cold assessment of the role of the United States in the early 1970's found receptive ears with Henry Kissinger, a devotee of unsentimental balance of power *Realpolitik*. Nixon had also given Kissinger the task of heading a top secret Government task force to examine the relation between population growth in developing nations, and its relation to US national security.

The motivation behind the secret task force had come from John D. Rockefeller and the Rockefeller Population Council. The core idea went back to the 1939 Council on Foreign Relations' War and Peace Studies Project leader, Isaiah Bowman. Global depopulation and food control were to become US strategic policy under Kissinger. This was to be the new "solution" to the threats to US global power and to its continued access to cheap raw materials from the developing world.

Notes

1. For a brief introduction to the extraordinary post-1945 bases of American global hegemony, useful are the following: Henry Luce, "The American Century", *Life*, 17 February 1941. New York Council on Foreign Relations, *The War & Peace Studies* summarized in http://www.cfr.org. Neil Smith, *American Empire: Roosevelt's Geographer and the Prelude to Globalization*, University of California Press, Berkeley, 2003. André Gunder Frank, *Crisis: In the World Economy*, Heinemann, London, 1980.

2. Francis J. Gavin, *Ideas, Power and the Politics of America's International Monetary Policy during the 1960's*, http://www.utexas.edu/lbj/faculty/gavin. See also F. William Engdahl, *A Century of War: Anglo-American Oil Politics and the New World Order*, Pluto Press Ltd, London, 2004, for a discussion of the de Gaulle gold issue. Also Central Intelligence Agency, Directorate of Intelligence, *French Actions and the Recent Gold Crisis*, Washington, D.C., 20 March 1968.

3. Samuel Huntington, et al., *The Crisis of Democracy: Report on the Governability of Democracies to the Trilateral Commission*, Trilateral Commission, New York University Press, 1975.

4. *Ibid.*

5. *Ibid.* Zbigniew Brzezinski, *Between Two Ages: America's Role in the Technotronic Era*, Harper Publishing House, New York, 1970.

6. Clifton B. Luttrell, *The Russian Wheat Deal—Hindsight vs. Foresight*, Federal Reserve Bank of St. Louis, October 1972, p. 2.

7. F. William Engdahl, *A Century of War: Anglo-American Oil Politics and the New World Order*, 2004, London, Pluto Press Ltd., pp.130-138.

8. Time, *What to Do: Costly Choices*, 11 November, 1974, p.6.

9. US Department of Agriculture, *World Grain Consumption and Stocks, 1960-2003*, Washington DC, Production, Supply & Distribution, Electronic Database, updated 9 April 2004.

10. Sen. George McGovern, cited in Laurence Simon, "The Ethics of Triage: A Perspective on the World Food Conference", *The Christian Century*, 1-8 January 1975.

11. Laurence Simon, *op. cit.*

12. *Ibid.* For a more complete discussion of the role of Kissinger in the 1973 oil price shock, see F. William Engdahl, *op. cit.*

13. Walter B. Saunders cited in A.V. Krebs, editor, *The Agribusiness Examiner*, Issue # 31, 26 April 1999.

14. A.V. Krebs, *op. cit.*

15. *J.W. Smith,* The World's Wasted Wealth 2, Institute for Economic Democracy, *1994, pp. 63, 64.*

16. A. V. Krebs, *Cargill & Co.'s* "Comparative Advantage in Free Trade", *The Agribusiness Examiner,* #31, 26 April 1999.

17. Michael Hudson, *Super Imperialism: The Origins and Fundamentals of US World Dominance,* London, Pluto Press Ltd., Second Edition, 2003 (originally published in 1972), pp. 229-235 for an excellent elaboration on the political workings of PL 480 under Kissinger. In hearings before the US Senate on the PL 480 legislation, Sen. Milton R. Young observed that US agricultural surpluses could be used as an instrument of foreign policy: "In my opinion, we have been blessed and not cursed with some surpluses. We are in the position of a nation with agricultural surpluses, when many other nations are starving. When we have such surpluses, we have adverse farm prices…This bill proposes for the first time, I think, a very feasible and sound method of trying to make our agricultural surpluses available to other nations of the world who are needy and in want of these supplies." (Cited in Congressional Research Service, 1979: 2).

18. Noah Zerbe, *Feeding the Famine? American Food Aid and the GMO Debate in Southern Africa,* Catholic University of Louvain, Belgium, in http://www.geoci ties.com/nzerbe/pubs/famine.pdf., pp. 9-10.

19. NACLA, "US Grain Arsenal" Chapter 2: «The Food Weapon: Mightier than Missiles.», *Latin America and Empire Report,* October 1975, http:// www.eco.utexas.edu/facstaff/Cleaver/357Lsum_s4_NACLA_Ch2.html.

20. *Ibid.*

21. George F. Kennan, "PPS/23: Review of Current Trends in U.S. Foreign Policy", *Foreign Relations of the United States,* 1948, Volume I, pp. 509-529. Policy Planning Staff Files, Memorandum by the Director of the Policy Planning Staff (Kennan)2 to the Secretary of State and the Under Secretary of State (Lovett). TOP SECRET.PPS/23. [Washington,] 24 February 1948. Kennan, one of the most influential shapers of the US Cold War, was author of a famous 1947 article in Foreign Affairs, the magazine of the New York Council on Foreign Relations. The article, *"The Sources of Soviet Conduct,"* was published in Foreign Affairs in July 1947. Signed under the pseudonym, "X" the true author was Kennan, who had been Ambassador Averell Harriman's Deputy in Moscow in 1946. The article set out the doctrine of containment of the Soviet Union, later known as the Cold War.

CHAPTER 4
A Secret US National Security Memo

> "Control oil and you control the nations; control food and you control the people…"
>
> *Henry Kissinger*

Population Growth and National Security

In April 1974, as a worldwide drought and the transformation of American farm policy was in full gear, Nixon's Secretary of State and National Security Adviser, Henry A. Kissinger, sent out a classified memo to select cabinet officials, including the Secretary of Defense, the Secretary of Agriculture, the Deputy Secretary of State and the CIA Director.

The title of the top secret memo was *Implications of Worldwide Population Growth for US Security and Overseas Interests.* The memo dealt with food policy, population growth and strategic raw materials. It had been commissioned by Nixon on the recommendation of John D. Rockefeller III. The secret project came to be called in Washington bureaucratic shorthand, NSSM 200, or National Security Study Memorandum 200.[1]

It was deemed that, should it ever be publicized or leaked, NSSM 200 would be explosive. It was kept secret for almost 15 years until private legal action by organizations associated with the Catholic Church finally forced its declassification in 1989. After a disgraced Nixon resigned over the Watergate scandal in 1975, his successor,

Gerald Ford, wasted no time in signing the Executive Order making NSSM 200 official US government policy.

The US decision to draft the policy came after the 1974 UN Population Conference in Bucharest, Romania, at which the UN failed to adopt the US position. That position had been shaped by the Rockefeller Foundation and most directly, by John D. Rockefeller III, and consisted in adopting a "world population plan of action" for drastic global population reduction policies. A fierce resistance from the Catholic Church, from every Communist country except Romania, as well as from Latin American and Asian nations, convinced leading US policy circles that covert means were needed to implement their project. It was entrusted to Henry Kissinger to draft that strategy, NSSM 200.

In his original initiating memo, Kissinger stated:

> The President has directed a study of the impact of world population growth on US security and overseas interests. The study should look forward at least until the year 2000, and use several alternative reasonable projections of population growth.
> - In terms of each projection, the study should assess:
> - the corresponding pace of development, especially in poorer countries;
> - the demand for US exports, especially of food, and the trade problems the US may face arising from competition for resources; and
> - the likelihood that population growth or imbalances will produce disruptive foreign policies and international instability.
> ...
> The study should focus on the international political and economic implications of population growth rather than its ecological, sociological or other aspects.
> The study would then offer possible courses of action for the United States in dealing with population matters abroad, particularly in developing countries, with special attention to these questions:
> - What, if any, new initiatives by the United States are needed to focus international attention on the population problem?
> - Can technological innovations or development reduce growth or ameliorate its effects?[2]

By December 1974, Kissinger had completed his policy document which included precise conclusions pertaining to global population growth:

> The most serious consequence for the short and middle term is the possibility of massive famines in certain parts of the world, especially the poorest regions. World needs for food rise by 2-1/2 percent or more per year ... at a time when readily available fertilizer and well-watered land is already largely being utilized. Therefore, additions to food production must come mainly from higher yields. Countries with large population growth cannot afford constantly growing imports, but for them to raise food output steadily by two to four percent over the next generation or two is a formidable challenge. Capital and foreign exchange requirements for intensive agriculture are heavy, and are aggravated by energy cost increases and fertilizer scarcities and price rises. The institutional, technical, and economic problems of transforming traditional agriculture are also very difficult to overcome.[3]

In December 1974, the world was in the early weeks of an oil price shock which saw oil prices explode by a staggering 400% within the coming six months, with profound consequences for world economic growth. Kissinger had personally played the key, behind-the-scenes role in manipulating that oil shock. He knew very well the impact that higher petroleum prices would have on world food supply. He was determined to use these higher oil prices to US strategic advantage.

Kissinger wrote in his NSSM report, referring to the poorer developing countries using the term, Least Developing Countries (LDCs):

> The world is increasingly dependent on mineral supplies from developing countries, and if rapid population growth frustrates their prospects for economic development and social progress, the resulting instability may undermine the conditions for expanded output and sustained flows of such resources.
>
> There will be serious problems for some of the poorest LDCs with rapid population growth. They will increasingly find it difficult to pay for needed raw materials and energy. Fertilizer, vital for

their own agricultural production, will be difficult to obtain for the next few years. Imports for fuel and other materials will cause grave problems which could impinge on the US, both through the need to supply greater financial support and in LDC efforts to obtain better terms of trade through higher prices for exports.

Economic Development and Population Growth
Rapid population growth creates a severe drag on rates of economic development otherwise attainable, sometimes to the point of preventing any increase in per capita incomes. In addition to the overall impact on per capita incomes, rapid population growth seriously affects a vast range of other aspects of the quality of life important to social and economic progress in the LDCs.[4]

The Washington blueprint was explicit. The United States should be in the forefront in promoting population reduction programs, both directly through the aid programs of the Government, making acceptance of birth reduction programs a prerequisite for US help. Or it should act indirectly, via the UN or the Bretton Woods institutions (IMF and World Bank).

Bluntly, the new US policy was to be, "if these inferior races get in the way of our securing ample, cheap raw materials, then we must find ways to get rid of them." This was the actual meaning of NSSM 200, in refined bureaucratic language.

Explicitly on population control, the NSSM 200 declared,

> [T]he US strategy should support general activities capable of achieving major breakthroughs in key problems which hinder attainment of fertility control objectives. For example, the development of more effective, simpler contraceptive methods through bio-medical research will benefit all countries which face the problem of rapid population growth; improvements in methods for measuring demographic changes will assist a number of LDCs in determining current population growth rates and evaluating the impact over time of population/family planning activities.[5]

Kissinger knew what he referred to when he spoke of "simpler contraceptive methods through bio-medical research." He was in close contact with the Rockefeller family and that wing of the US establishment which promoted bio-medical research as a new form of population control. Auschwitz revelations regarding its use had made the term unsavoury. Before World War II, it was known as eugenics. It was renamed by its promoters the more euphemistic "population control" after the war. The content was unchanged: reduce "inferior" races and populations in order to preserve the control by "superior" races.

Food for Cargill & Co.

The NSSM 200 also bore the strong mark of William Pearce and the Cargill agribusiness trade lobby. In a section titled, "Food for Peace and Population," Kissinger wrote, "One of the most fundamental aspects of the impact of population growth on the political and economic well-being of the globe is its relationship to food. Here the problem of the interrelationship of population, national resources, environment, productivity and political and economic stability come together when shortages of this basic human need occur."[6]

He continued, "The major challenge will be to increase food production in the LDCs themselves, and to liberalize the system in which grain is transferred commercially from producer to consumer countries."

In effect, he proposed spreading the Rockefeller Foundation's Green Revolution while also demanding removal of protective national trade barriers. The objective was to open the way for a flood of US grain imports in key developing markets. Explicitly, Kissinger proposed, "Expansion of production of the input elements of food production (i.e., fertilizer, availability of water and high yield seed stocks) and increased incentives for expanded agricultural productivity"—the essence of the Green Revolution. It went without saying that US agribusiness companies would supply the needed fertilizer and special high-yield seeds. That was what the so-called Green Revolution had really been about in the 1960's.

NSSM 200 called for, "New international trade arrangements for agricultural products, open enough to permit maximum production by efficient producers ...," not coincidentally, just the demand of Cargill, ADM, Continental Grain, Bunge and the giant agribusiness corporations then emerging as major US nationally strategic corporations.

The NSSM document packaged the earlier Kissinger "food as a weapon" policy in new clothes:

> Food is another special concern in any population strategy. Adequate food stocks need to be created to provide for periods of severe shortages and LDC food production efforts must be re-enforced to meet increased demand resulting from population and income growth. US agricultural production goals should take account of the normal import requirements of LDCs (as well as developed countries) and of likely occasional crop failures in major parts of the LDC world. Without improved food security, there will be pressure leading to possible conflict and the desire for large families for «insurance» purposes, thus undermining ... population control efforts.
>
> ...
>
> [T]o maximize progress toward population stability, primary emphasis would be placed on the largest and fastest growing developing countries where the imbalance between growing numbers and development potential most seriously risks instability, unrest, and international tensions. *These countries are: India, Bangladesh, Pakistan, Nigeria, Mexico, Indonesia, Brazil, The Philippines, Thailand, Egypt, Turkey, Ethiopia, and Colombia .... This group of priority countries* includes some with virtually no government interest in family planning and others with active government family planning programs which require and would welcome enlarged technical and financial assistance. These countries should be given the highest priority within AID's population program in terms of resource allocations and/or leadership efforts to encourage action by other donors and organizations.[7]

The Unlucky Thirteen...

Thirteen developing countries, including India, Nigeria, Mexico, Indonesia, Brazil, Turkey, and Colombia, encompassed some of the most resource-rich areas on the planet. Over the following three decades they were also to be among the most politically unstable. The NSSM 200 policy argued that only a drastic reduction in their populations would allow US exploitation of their raw materials.

Naturally, Kissinger knew that if it were be revealed that the US Government was actively promoting population reduction in raw materials-rich developing countries, Washington would be accused of imperialist ambitions, genocide and worse. He proposed a slick propaganda campaign to hide this aspect of NSSM 200:

> The US can help to minimize charges of an imperialist motivation behind its support of population activities by repeatedly asserting that such support derives from a concern with:
> a) the right of the individual couple to determine freely and responsibly their number and spacing of children and to have information, education, and means to do so; and
> b) the fundamental social and economic development of poor countries in which rapid population growth is both a contributing cause and a consequence of widespread poverty.
> Furthermore, the US should also take steps to convey the message that the control of world population growth is in the mutual interest of the developed and developing countries alike.[8]

In so many words, population control on a global scale was now to be called, "freedom of choice," and "sustainable development." George Orwell could not have done better. The language had been lifted from an earlier Report to President Nixon from John D. Rockefeller III.

NSSM 200 noted that the volume of grain imports needed by developing countries would "grow significantly." It called for trade liberalization in grain imports around the world to address this alleged problem, a "free market" not unlike the one Britain demanded when its manufactured goods dominated world markets after the Corn Laws repeal in 1846.

Like the "population bomb," the food crisis was also a manufactured hype in the 1970's, a hype helped by the sudden oil price shock on developing economies. The image of vast areas of the world, teeming with "overpopulation" and riotings or killings, were run repeatedly on American TV to drive the point home. In reality, the "problems" in developing sector agriculture were mainly that it didn't offer enough opportunities for the major US agribusiness companies. Cargill and the giant US grain trading companies were not far away from Kissinger's door.

The NSSM report added that, "The location of known reserves of higher-grade ores of most minerals favors increasing dependence of all industrialized regions on imports from less developed countries. The real problems of mineral supplies lie, not in basic physical sufficiency, but in the politico-economic issues of access, terms for exploration and exploitation, and division of the benefits among producers, consumers, and host country governments." Forced population control programs and other measures were to be deployed if necessary, to ensure US access to such strategic raw materials.

The document concluded, "In the longer run, LDCs must both decrease population growth and increase agricultural production significantly."

While arguing for reducing global population growth by 500 million people by the year 2000, curiously enough, Kissinger noted elsewhere in his report that the population problem was already causing 10 million deaths yearly. In short he advocated doubling the death rate to at least 20 million, in the name of addressing the problem of deaths due to lack of sufficient food. The public would be led to believe that the new policy, at least what would be made public, was a positive one. In the strict definition of the UN Convention of 1948, it was genocide.

Kissinger went on to suggest the kinds of coercive measures the US policy elite now envisioned. He bluntly stated that food aid should be considered, "an instrument of national power." Then, in a stark comment, he suggested *the US would ration its food aid to "help people who can't or won't control their population growth."* (emphasis added). Sterilize or starve ... It was little wonder the document was classified "Top Secret."

NSSM 200 was remarkable in many respects. It made depopulation in foreign developing countries an explicit, if secret, strategic national security priority of the United States Government for the first time. It outlined what was to become a strategy to promote fertility control under the rubric "family planning," and it linked the population growth issue to the availability of strategic minerals.

However, one of the most significant aspects of NSSM 200 was that it reflected an emerging consensus with some of America's wealthiest families, its most influential establishment.

Kissinger was, in effect, a hired hand within the Government, but not hired by a mere President of the United States. He was hired to act and negotiate on behalf of the most powerful family within the postwar US establishment at the time—the Rockefellers.

In 1955 Nelson Rockefeller had invited Kissinger to become a study director for the Council on Foreign Relations. One year later, Kissinger became Director of the Special Studies Project for the Rockefeller Brothers Fund, where he came to know the family on a first-name basis. Kissinger later married a Rockefeller employee, Nancy Maginnes, to round the connection.

By November 1975, Richard Nixon had been forced from office in the mysterious Watergate affair, some suspected on the machinations of a politically ambitious Nelson Rockefeller, working with Kissinger and Alexander Haig. Nixon's successor, a non-descript Gerald Ford, appointed Nelson Rockefeller to be his Vice President. Nelson was in effect, "a heartbeat away" from his dream of being President. Nelson's old friend Kissinger was Secretary of State.

On November 1975, President Ford signed off on Kissinger's NSSM 200 as official US foreign policy. Kissinger had been replaced by his assistant and later business partner, Brent Scowcroft, as NSC head. Scowcroft dutifully submitted Kissinger's NSSM 200 draft, to the new President for signature. Kissinger remained Secretary of State and Nelson Rockefeller, Vice President. The US was going into the depopulation business big time, and food control was to play a central role in that business.

Brazil as NSSM 200 "Model"
The secret Kissinger plan was implemented immediately. The thirteen priority countries for population reduction were to undergo drastic changes in their affairs over the following thirty years. Most would not even be aware of what was happening.

Brazil was one of the most clearly documented examples of NSSM 200. Beginning in the late 1980's, almost 14 years into the implementation of NSSM 200, the Brazilian Ministry of Health began to investigate reports of massive sterilization of Brazilian women. The government investigation was the result of a formal Congressional inquiry, sponsored by more than 165 legislators from every political party represented in the Brazilian legislature.[9]

The investigation had been initiated after information about the secret US National Security Council memorandum on American population control objectives in developing countries was published in the *Jornal de Brasilia, Hova do Povo* (Rio de Janeiro), *Jornal do Brasil,* and other major Brazilian newspapers in May 1991.

The Brazilian government was shocked to find that an estimated 44% of all Brazilian women aged between 14 and 55 had been permanently sterilized. Most of the older women had been sterilized when the program started in the mid-1970's. The Government found that the sterilizations had been carried out by a variety of different organizations and agencies, some Brazilian. They included the International Planned Parenthood Federation, the US Pathfinder Fund, the Association for Voluntary Surgical Contraception, Family Health International—all programs under the aegis and guidance of the US Agency for International Development (USAID).[10]

By 1989, the Brazilian government, which initially had been convinced to cooperate in the interest of economic growth and poverty alleviation, protested to USAID that the sterilization programs had become "overwhelming and unnecessary." According to some reports, under the program, as many as 90% of all Brazilian women of African descent had been sterilized, which would eliminate future generations in a nation whose Black population is second only to Nigeria's. Almost half of Brazil's 154 million people in the 1980's were believed to be of African ancestry.[11]

Kissinger in NSSM 200 had noted the special role of Brazil. It was on the target list of thirteen countries because "it clearly dominates the Continent [South America] demographically," and its population would be projected to equal that of the United States by the year 2000. Such growth from Brazil, the NSSM memo warned, implied "a growing power status for Brazil in Latin America and on the world scene over the next 25 years."[12]

Behind Kissinger, Scowcroft and the assorted Washington civil servants who carried out the new NSSM 200 policy, stood a circle of private, enormously influential persons. None were more influential at the time than the Rockefeller brothers. On population policy, no Rockefeller held more clout than John D. Rockefeller III, grandson of the Standard Oil founder.

John D. Rockefeller III was appointed by President Nixon in July 1969 to head the Commission on Population Growth and the American Future. Their report prepared the ground for Kissinger's NSSM 200. In 1972, some months before Kissinger's secret project began, Rockefeller presented his report to the President. In the election year, Nixon decided to downplay the report and, as a result, it got little press attention. Its policy recommendations, however, received major priority. Rockefeller proposed what were then drastic measures to stem an alleged population explosion in the United States.[13]

Among his recommendations were the establishment of sex education programs in all schools, population education so that the public appreciated the supposed crisis, and the repeal of all laws that hindered contraceptive means to minors and adults. It proposed making voluntary sterilization easier and liberalizing state laws against abortion. Abortion had been regarded as a major vehicle for fertility control by the Rockefeller circles for decades, hindered by strong opposition from church and other groups.

What came next under NSSM 200 could only be understood from the vantage point of the background to John D. Rockefeller III's obsession with population growth. Henry Kissinger's National Security Council NSSM 200 paper on population control (1974) expressed the assumptions of a decades old effort to breed human traits, known until the end of the Third Reich as Eugenics.

Notes

1. Henry Kissinger, *National Security Study Memorandum 200, April 24, 1974: Implications of Worldwide Population Growth for US Security and Overseas Interests,* Initiating Memo. Complete text is contained in Stephen Mumford, *The NSSM 200 Directive and The Study Requested,* 1996, http://www.population-security.org/11-CH3.html.

2. *Ibid.* According to the magazine Catholic World Reporter, "The key document needed to understand U.S. policy toward world population during the past 20 years … was declassified in 1980 but not made publicly available until June 1990. Dated December 10, 1974, it is a study by the National Security Council (NSC) entitled «NSSM 200: Implications of Worldwide Population Growth for U.S. Security and Overseas Interests." This document views population growth in less developed countries as not only a serious threat to strategic interests of the U.S. but also as the prime cause of political instability in Third World nations, threatening American overseas investments.

3. *Ibid.,* "Adequacy of World Food Supplies", *Executive Summary,* paragraph 6.

4. *Ibid., Executive Summary,* paragraphs 9-10.

5. *Ibid.,* Part II: Policy Recommendations, II. *Action to Create Conditions for Fertility Decline §3. Mode and Content of U.S. Population Assistance.*

6. *Ibid.,* Part II. Policy Recommendations: *C. Food for Peace Program and Population,Discussion.*

7. *Ibid.,* Part II: Policy Recommendations, I. Introduction—A U.S. Global Population Strategy, B. Key Country priorities in U.S. and Multilateral Population Assistance. (Emphasis added).

8. *Ibid.,* Part II: Policy Recommendations, I. *Introduction—A U.S. Global Population Strategy, F. Development of World-Wide Political and Popular Commitment to Population Stabilization and Its Associated Improvement of Individual Quality of Life.*

9. Andre Caetano, *Fertility Transition and the Transition of Female Sterilization in Northeastern Brazil: The Roles of Medicine and Politics,* http://www.iussp.org/Brazil2001/s10/S19_02_Caetona.pdf. p. 19. Details of the Brazilian Congress inquiry are in Baobab Press, *Brazil Launches Inquiry into US Population Activities,* Vol. 1, no. 12, Washington D.C., http://india.indymedia.org/en/2003/05/4869.html. Alternate location in http://thepragmaticprogressive.blogspot.com/2003/05/this-article-printed-in-its-entirety.html.

10. *United Nations Population Fund Inventory of Population Projects in Developing Countries Around the World,* cited in Baobab Press, *op. cit.*

11. Baobab Press, *op. cit.*

12. Henry Kissinger, *op. cit.*, "Part One: Analytical section", *Chapter I—Highlight of World Demographic Trends: Latin America.*

13. John D.Rockefeller III, Report of the Commission on Population Growth and the American Future, Washington, D.C. 27 March 1972.

CHAPTER 5

The Brotherhood of Death

Human Guinea Pigs

Years before Henry Kissinger and Brent Scowcroft made population reduction the official foreign policy of the United States Government, the Rockefeller brothers, in particular John D. Rockefeller III, or JDR III as he was affectionately known, were busy experimenting on human guinea pigs.

In the 1950's, brother Nelson Rockefeller had invested in exploiting the cheap, non-unionized labor of Puerto Ricans in the New York garment center sweatshops, flying them into New York at cheap rates on the family's Eastern Airlines shuttle. He was also engaged in setting up cheap labor manufacturing directly on the island, far away from prying US health and industrial safety regulators, under a government program named Operation Bootstrap. Operation Bootstrap was launched in 1947 to offer US firms the benefit of a cheap labor force as well as generous tax holidays.[1]

At the time, Nelson Rockefeller was Under Secretary of the Department of Health, Education, and Welfare and a shadowy and highly influential figure in the Eisenhower Administration.

In Nelson's version of Operation Bootstrap, the boots were owned by the Rockefeller family and their business friends around David Rockefeller's Chase Bank. Chase's most profitable business during the 1950's was via Puerto Rico and Operation Bootstrap, financing runaway sweatshops fleeing higher wages in the US. The family-controlled company, International Basic Economy Corporation (IBEC) built up vast assets on the island.[2] The only straps were those used by sweatshop owners on the island to force a higher level of productivity from their workers.

While Nelson was busy encouraging the spirit of free enterprise among Puerto Ricans, brother John D. III was running human experiments in mass sterilization on the poorer citizens of Puerto Rico. Puerto Rico was an unfortunate island whose sovereignty got lost somewhere in the shuffle of American diplomacy. It was a de facto US colony, with ultimate legal control decided in Washington, making it an ideal experiment station. Through his newly founded Population Council, JDR III first ran some of the experiments in population reduction which would later become global State Department policy under Henry Kissinger's NSSM 200.[3]

JDR III made Puerto Rico into a huge laboratory to test his ideas on mass population control beginning in the 1950's. By 1965, an estimated 35% of Puerto Rico's women of child-bearing age had been permanently sterilized, according to a study made that year by the island's Public Health Department.[4] The Rockefeller's Population Council, and the US Government Department of Health Education and Welfare—where brother Nelson was Under-Secretary—packaged the sterilization campaign. They used the spurious argument that it would protect women's health and stabilize incomes if there were fewer mouths to feed.

Poor Puerto Rican women were encouraged to give birth in sanitary new US-built hospitals where doctors were under orders to sterilize mothers who had given birth to two children by tying their tubes, usually without the mothers' consent. By 1965, Puerto Rico was a world leader in at least one category. It had the highest percentage of sterilized women in the world. India lagged badly in comparison, with a mere 3%. It made a difference when the

Rockefeller family could control the process directly without government meddling.[5]

"Second Only to Control of Atomic Weapons…"

John D. III's forced sterilization program was no radical departure for the family. The Rockefellers had long regarded Puerto Rico as a convenient human laboratory. In 1931, the Rockefeller Institute for Medical Research, later renamed the Rockefeller University, financed the cancer experiments of Dr. Cornelius Rhoads in Puerto Rico.

Rhoads was no ordinary scientist. It later came out that Rhoads had deliberately infected his subjects with cancer cells to see what would happen. Eight of his subjects died. According to pathologist Cornelius Rhoads, "Porto Ricans are beyond doubt the dirtiest, laziest, most degenerate and thievish race of men ever inhabiting this sphere. What the island needs is not public health work but a tidal wave or something to totally exterminate the population. I have done my best to further the process of extermination by killing off 8 …"[6] Initially written in a confidential letter to a fellow researcher, Rhoads's boast of killing Puerto Ricans appeared in *Time* magazine in February 1932 after Pedro Albizu Campos, leader of the Puerto Rican Nationalist Party, gained possession of the letter and publicized its contents.[7]

Rather than being tried for murder, the Rockefeller Institute scientist was asked to establish the US Army Biological Warfare facilities in Maryland, Utah and also Panama, and was later named to the US Atomic Energy Commission, where radiation experiments were secretly conducted on prisoners, hospital patients and US soldiers.[8]

In 1961, more than a decade before his policies were to become enshrined in NSSM 200, JDR III gave the Second McDougall Lecture to the United Nations Food and Agriculture Organization. Rockefeller told the listeners, "To my mind, population growth is second only to control of atomic weapons as the paramount problem of the day." He spoke of a "cold inevitability, a certainty that is mathematical, that gives the problems posed by too-rapid population growth a somber and chilling caste indeed." The "grim fact" of population growth, he warned, "cuts across all the basic needs of mankind and ... frustrates man's achievement of his higher needs."[9]

Rockefeller Supports Eugenics

JDR III grew up surrounded by eugenicists, race theorists and Malthusians at the Rockefeller Foundation such as Frederick Osborn, Henry Fairchild and Alan Gregg. For John D. III, it seemed only natural that he and others of his "class" should decide which elements of the human species survived, in order that they could have "life as we want it to be." They saw it as being a bit like culling herds of sheep for the best of breed.

The logic of human life for the family was simple: supply and demand. As Jameson Taylor expressed it,

> For Rockefeller, the proper care of sheep ... requires nothing more than an equalization of supply with demand. If supply—i.e., food, water and space—cannot meet demand, supply must be increased and demand must be decreased. The Rockefeller Foundation has used this two pronged approach to great effect. The supply shortage has been addressed by ... advanced medical practices and increased crop yields. The demand problem has been solved by culling the herd via birth control and abortion.[10]

For most Americans and for most of the world, the idea that the leading policy circles of the United States Government, acting on the behest of some of its wealthiest families and most influential universities, would deliberately promote the mass covert sterilization of entire population groups was too far-fetched to accept.

Few realized that individuals with names such as Rockefeller, Harriman, banker J.P. Morgan Jr., Mary Duke Biddle of the tobacco family, Cleveland Dodge, John Harvey Kellogg from the breakfast cereal fortune, Clarence Gamble of Proctor & Gamble, were quietly funding eugenics as members of the American Eugenics Society. They had also been financing experiments in forced sterilization of "inferior people" and various forms of population control as early as World War I. Their counterparts in the English Eugenics Society at the time included the British Chancellor of the Exchequer, Winston Churchill, economist John Maynard Keynes, Arthur Lord Balfour and Julian Huxley, who went on to be the first head of UNESCO after the war.

Combating "The Human Cancer"

Population and related food policies of the US Government of the early 1970's emanated from the halls of the Rockefeller Foundation, from their Population Council and the Rockefeller Brothers Fund, and from a handful of similarly well-endowed private foundations, such as the Ford Foundation and the Carnegie Foundation. The true history of those organizations was carefully buried behind a façade of philanthropy. In reality, these tax-exempt foundations served as vehicles for the advancement and domination of powerful elite families at the expense of the welfare of most American citizens and of most of mankind.

One man served as head of the Medical Division of the Rockefeller Foundation for more than 34 years. His name was Alan Gregg. An all-but unknown person to the outside world, throughout his 34 years at the Medical Division of the Rockefeller Foundation, Gregg wielded tremendous influence. He was Vice President of the Foundation on his retirement in 1956, and his ideology pervaded the institute decades after. It was an ideology of Malthusian brutality and racist finality.

Gregg once wrote in an article for a scientific journal on population, "There is an alarming parallel between the growth of a cancer in the body of an organism and the growth of human population in the earth ecological economy." He then asserted that "cancerous growths demand food, but so far as I know, they have never been cured by getting it. The analogies can be found in our plundered planet."[11]

This was a formulation which translated as, "people pollute, so eliminate pollution by eliminating people..." Gregg then went on, in a paper commissioned by *Science*, one of the most eminent scientific journals in the US, to observe, "how nearly the slums of our great cities resemble the necrosis of tumors." And this "raised the whimsical query: Which is the more offensive to decency and beauty, slums or the fetid detritus of a growing tumor?"[12]

The Rockefellers' Darker Secrets
The role of the Rockefeller Foundation in US and global population policy was not accidental, nor was it a minor aspect of the institution's mission. It was at the very heart of it. This population policy role was the key to understanding the later engagement of the Foundation in the revolution in biotechnology and plant genetics.

In 1913, the founder of the Standard Oil trust, John D. Rockefeller Sr., was advised to hide his wealth behind a tax-exempt foundation. That year Congress had passed the first federal income tax, and the Rockefeller family and other wealthy Americans such as steel magnate, Andrew Carnegie, were enraged at what they deemed illegal theft of justly-earned gains. As Carnegie put it at the time, "Wealth passing through the hands of the few can be made a much more potent force for the elevation of our race (sic) than if distributed in small sums to the people themselves."[13] In so many words, money should only belong to the very wealthy, who know best how to use it.

The newly-established Rockefeller Foundation's stated mission was, "to promote the wellbeing of mankind throughout the world." It went without saying that the foundation alone, and the Rockefeller family, would decide just what "promoting the wellbeing of mankind" entailed.

From its inception, the Rockefeller Foundation was focused on culling the herd, or systematically reducing populations of "inferior" breeds. One of the first Rockefeller Foundation grants was to the Social Science Research Council for study of birth control techniques in 1923. In 1936, the foundation created and endowed the first Office of Population Research at Princeton University, headed by Eugenics Society member Frank Notestein, to study the political aspects of population change.

From its founding onwards, the philosophy of the Rockefeller Foundation was to deal with "causes rather than symptoms." Clearly one of the "causes" of world problems, as the family saw it, was the persistent tendency of the human species, at least the less wealthy portion of it, to reproduce and multiply itself. An increasing number of people in the world meant a greater potential to cause trouble and to demand a bigger slice of the Big Pie of Life, which the Rockefellers and their wealthy friends regarded exclusively as their "God-given" right.

Back in 1894, when the family's oil fortune was in its early days, JDR III's father, John D. Junior, wrote an essay as a student at Brown University entitled, "The Dangers to America Arising from Unrestricted Immigration." In it he wrote about immigrants, then mostly from Italy, Ireland and the rest of Europe, calling them, "the scum of foreign cities, the vagabond, the tramp, the pauper, and the indolent ... ignorant and hardly better than beasts."[14]

"The Best of Breed"—Eugenics and the "Master Race"

One of the first philanthropic projects undertaken by the Rockefeller Foundation in the 1920's was to fund the American Eugenics Society, and the Eugenics Record Office at Cold Spring Harbor, New York where by 1917, John D. Rockefeller had become the office's second largest supporter after the Harriman family.

Eugenics was a pseudo-science. The word was first coined in England in 1883 by Charles Darwin's cousin, Francis Galton, and it was founded on Darwin's 1859 work, the *Origin of Species.* Darwin had imposed what he termed, "the application of the theories of Malthus to the entire vegetable and animal kingdom." Malthus, who shortly before his death had repudiated his own theory of population, asserted in his 1798 tract, *Essay on the Principles of Population,* that populations tend to expand geometrically while food supply grew only arithmetically, leading to periodic famine and death to eliminate the "surplus" populations.

During the latter 19th Century, an explosion of population in Europe and North America was accompanied, thanks to the application of science and technological improvements, by rising living standards and increased food supply, thus discrediting Malthusianism as a serious science. However, by the 1920's, Rockefeller,

Carnegie and other vastly wealthy Americans embraced a Malthusian notion of what came to be called, "social Darwinism," which justified their accumulation of vast fortunes with the argument that it was a kind of divine proof of their superior species' survival traits over less fortunate mortals.

A related major Rockefeller Foundation project in the 1920's was the financing of Margaret Sanger's Planned Parenthood Federation of America, initially known as the American Birth Control League, a racist association promoting eugenics in the form of population control and forced sterilization, under the guise of rational "family planning." She wrote: "Birth control is thus the entering wedge for the Eugenic educator ... the unbalance between the birth rate of the 'unfit' and the 'fit' is admittedly the greatest present menace to civilization."[15]

Sanger, portrayed as a selfless woman of charity, was in reality a committed eugenicist, an outright race supremacist, who remained a Rockefeller family intimate until her death. She railed against "inferior classes" and was obsessed with "how to limit and discourage the over-fertility (sic) of the mentally and physically defective."[16]

As it was defined by its sponsors, eugenics was the study of improving the "quality" of the human species, while reducing the quantity of "inferior beings," or as Sanger put it, the "qualitative factor over the quantitative factor ... in dealing with the great masses of humanity." The title page of the Eugenics Review, the journal of the Eugenics Education Society, carried the original definition of British eugenics founder, Francis Galton, Darwin's cousin, who defined eugenics as "the science of improvement of the human race germ plasm through better breeding. Eugenics is the study of agencies under social control that may improve or impair the racial qualities of future generations, whether physically or mentally."

In her 1922 book, The Pivot of Civilization, in which among other proposals she advocates the idea of parenthood licenses—no one being permitted to have a child unless they first obtain a government-approved parenthood permit, Sanger wrote, "Birth control ... is really the greatest and most truly eugenic program and its adoption as part of the program of Eugenics would immediately give a concrete and realistic power to that science ... as the most constructive and necessary of the means to racial health."[17] Margaret Sanger was appreciated in international circles for her population control zeal. In 1933, the head of the Nazi Physicians' Association, Reichsärzteführer, Dr. Gerhard Wagner, praised Sanger for her stringent racial policies asking fellow Germans to follow her model.

Contrary to popular belief, the idea of a Nordic master race was not solely a Nazi Germany fantasy. It had its early roots in the United States of America going back to the early years of the 20th Century.

The President of the prestigious Stanford University in California, David Starr Jordan, promoted the idea of "race and blood" in his 1902 book, "Blood of a Nation." He claimed that poverty was an inherited genetic trait, as was talent. Education had no influence; you either "had it" or you didn't.

Two years later, in 1904 Andrew Carnegie's Carnegie Institute had founded the major laboratory at Cold Spring Harbor, the Eugenics Record Office on wealthy Long Island, outside New York City, where millions of index cards on the bloodlines of ordinary Americans were gathered, to plan the possible removal of entire bloodlines deemed inferior. The land for the institute was donated by railroad magnate, E. H. Harriman, a firm supporter of eugenics. This was eugenics, American elite style. Naturally, if the ideal was tall, blond, blue-eyed Nordic types, that meant dark-skinned Asians, Indians, Blacks, Hispanics and others, including the sick and retarded, were deemed inferior to the eugenics goal of "best of breed."[18]

The aim of the index card project was to map the inferior bloodlines and subject them to lifelong segregation and sterilization to "kill their bloodlines." The sponsors were out to eliminate those they deemed "unfit." As early as 1911, Carnegie was funding an American Breeder's Association study on the "Best Practical Means for Cutting off the Defective Germ-Plasm in the Human Population."[19]

One of the largest and most significant financial contributors for various eugenics projects soon became the Rockefeller Foundation. It poured hundreds of thousands of dollars into various eugenics and population projects, from the American Eugenics Society to Cold Spring Harbor, to the American Breeder's Association.[20]

One of the more prominent members of the American Eugenics Society in the early 1920's was Dr. Paul Bowman Popenoe, a US Army venereal disease specialist from World War I, who wrote a textbook entitled, "*Applied Eugenics.*" In sum, Popenoe said, "The first method which presents itself is execution ... Its value in keeping up the standard of the race should not be underestimated."[21] He went on to eloquently advocate the "destruction of the individual by some adverse feature of the environment, such as excessive cold, or bacteria or by bodily deficiency." In his book, Popenoe spoke of estimated five million Americans who would, at one point or another, end up in mental hospitals, and of "five million more who are so deficient intellectually with less than 70% average intelligence, as to be in many cases, liabilities rather than assets to the race."[22] The book was aimed at a select, elite readership. It was an example of what the eugenics movement termed "negative eugenics"—the systematic elimination of "inferior" beings, whether mentally inferior, physically handicapped or racially non-white.

Popenoe's radical approach was a bit too controversial for some, but by 1927, in Buck vs. Bell, the US Supreme Court, in a decision by Justice Oliver Wendell Holmes, ruled that the forced sterilization program of the State of Virginia was Constitutional. In his written decision, Holmes wrote, "It is better for all the world, if instead of waiting to execute degenerate offspring for crime, or to let them starve for their imbecility, society can prevent those who are manifestly unfit from continuing their kind.... Three generations of imbeciles are enough."[23] Holmes, one of the most influential Supreme Court justices, was also one of its most outspoken racists. In 1922, the ageing Holmes wrote to British economist and Labour Party leading figure, Harold J. Laski, "As I have said, no doubt often, it seems to me that all society rests on the death of men. If you don't kill 'em one way you kill 'em another—or prevent their being born. Is not the present time an illustration of Malthus?" The statement could have served as the guiding slogan of the Rockefeller Foundation eugenics efforts.[24]

This 1927 Supreme Court decision opened the floodgates for thousands of American citizens to be coercively sterilized or otherwise persecuted as subhuman. One Illinois mental hospital in Lincoln fed new patients milk from known tubercular cows, in the conviction that a genetically strong human specimen would be immune.[25] The State of California was the eugenics model state. Under its sweeping eugenics law, passed in 1909, all feebleminded or other mental patients were sterilized before discharge, and any criminal found guilty of any crime three times could be sterilized at the discretion of a consulting physician. California sterilized some 9,782 individuals, mostly women classified as "bad girls," many of whom had been forced into prostitution.[26]

Years later, the Nazis at the Nuremberg trials quoted Holmes' words in their own defense. In the postwar world, not surprisingly, it was to no avail. The Rockefeller propaganda machine buried the reference; the victors defined the terms of peace and the truth of war.

"Calling a Spade a Spade..."

The Rockefeller enthusiasm for eugenics during the 1920's did not stop at America's own shores. Rockefeller Foundation money played an instrumental role in financing German eugenics during the 1920's. From 1922 to 1926, the Rockefeller Foundation donated through its Paris office a staggering $410,000 to a total of hundreds of German eugenics researchers. In 1926, it awarded an impressive $250,000 for the creation of the Berlin Kaiser Wilhelm Institute for Psychiatry. That was the equivalent of some $26 millions in 2004 dollars, a sum especially unheard of in a Germany devastated by Weimar hyperinflation and economic depression. During the 1920's Rockefeller Foundation money dominated and steered German eugenics research.[27]

As American researcher Edwin Black and others later documented, the leading psychiatrist at the Kaiser Wilhelm Institute at that time was Ernst Rüdin, a man who went on to a stellar career as the architect of Adolf Hitler's systematic program of medical eugenics. The Rockefeller-financed Rüdin was named President of the World Eugenics Federation in 1932. Their platform was openly advocating the killing or sterilization of people whose heredity made them a "public burden."

The Rockefeller Foundation largesse for German research was apparently unbounded in those days. In 1929, the year of the great Wall Street crash and extreme German economic crisis, Rockefeller gave a grant of $317,000 to the Kaiser Wilhelm Institute for Brain Research, one of several subsequent Rockefeller grants.[28]

The multi-talented Rüdin was also head of brain research at that institute, where Hermann J. Muller, an American eugenicist also funded with Rockefeller money, was employed. Later it was revealed that the institute received "brains in batches of 150-250" from victims of the Nazi euthanasia program at the Brandenburg State Hospital in the late 1930's.[29] The brain research was directed towards the Nazi experiments on Jews, gypsies, the mentally handicapped and other "defectives." In 1931, the Rockefeller Foundation approved a further ten-year grant of a sizeable $89,000 to Rüdin's Institute for Psychiatry to research links between blood, neurology and mental illness. Rockefeller money was funding eugenics in its purest form.[30]

Rüdin also led the Nazi program of forced eugenic sterilization, and was a prime architect of the 1933 Nazi Sterilization Law. Rudin and his staff, as part of the Task Force of Heredity Experts, chaired by SS chief Heinrich Himmler, drew up the sterilization law. Described as an "American Model" law, it was adopted in July 1933 and proudly printed in the September 1933 *Eugenical News* (USA), with Hitler's signature.[31] Rüdin called for sterilizing all members of an unfit individual's extended family. Rüdin was twice honoured by Adolf Hitler for his contribution to German eugenics and racial cleansing. Under his Sterilization Law, some 400,000 Germans were diagnosed as manic-depressive or schizophrenic and forcibly sterilized, and thousands of handicapped children were simply killed.[32] Declaring racial hygiene a "spiritual movement," Rüdin and his associates found a willing collaborator in Adolf Hitler. "Only through [the Führer] did our dream of over thirty years, that of applying racial hygiene to society, become a reality," Rüdin said.[33]

Hitler personally was a great enthusiast of American eugenics, praising US eugenics efforts in 1924 in *Mein Kampf:* "There is today one state in which at least weak beginnings toward a better conception of immigration are noticeable. Of course, it is not our model German Republic, but the United States."[34] A few years later, Hitler wrote the American eugenicist Madison Grant to personally praise his 1916 book, *The Passing of the Great Race.* In it Grant had written among other things that America had been "infested by a large and increasing number of the weak, the broken and the mentally crippled of all races . . . ." Grant advocated as a eugenic remedy "a rigid system of selection through the elimination of those who are weak or unfit—in other words, social failures (*sic*)."[35] Hitler clearly recognized a kindred soul in Grant, a co-founder of the American Eugenics Society.

By 1940, thousands of Germans from old age homes and mental institutions had been systematically gassed, as had been advocated twenty years earlier in the United States by Popenoe, if with limited success. In 1940, just back from a tour of the German eugenics institutes, Leon Whitney, Executive Secretary of the Rockefeller-funded American Eugenics Society, declared of the Nazi experiments, "While we were pussy-footing around . . . the Germans were calling a spade a spade."[36]

In May 1932, the Rockefeller Foundation sent a telegram to its Paris office, which quietly funnelled the US Rockefeller funds into Germany. The telegram read: "JUNE MEETING EXECUTIVE COMMITTEE: NINE THOUSAND DOLLARS OVER THREE YEAR PERIOD TO KWG INSTITUTE ANTHROPOLOGY FOR RESEARCH ON TWINS AND EFFECTS ON LATER GENERATIONS OF SUBSTANCES TOXIC FOR GERM PLASM."[37]

That was one year before Hitler became Chancellor. The "KWG" was the Kaiser Wilhelm Institute for Anthropology, Human Heredity and Eugenics in Berlin. The germ-plasm research was to continue well into the Third Reich, financed with Rockefeller Foundation money until at least 1939.[38]

The head of the German eugenics institute in Berlin was Otmar Freiherr von Verschuer. His research on twins had long been a dream of American eugenics advocates in order to advance their theories of heredity. In 1942, in the German Nazi eugenics journal *Der Erbarzt* which he edited, von Verschuer advocated a "total solution to the Jewish problem." In 1936, still receiving Rockefeller funds, Verschuer was called to Frankfurt to head a newly established Institute of Genetics and Racial Hygiene at the University of Frankfurt. The largest of its kind, the Frankfurt institute was responsible for the compulsory medical curriculum on eugenics and racial hygiene.[39]

Von Verschuer's long-time assistant was Dr. Josef Mengele, who headed human experiments at the Auschwitz concentration camp after May 1943. Von Verschuer was delighted when Mengele, who won the name "The Angel of Death" for his deadly experiments on human prisoners, was assigned to Auschwitz. Now their "scientific" research could continue uninhibited. He wrote at the time to the German Research Society that: "My assistant, Dr. Josef Mengele (M.D., Ph.D.) joined me in this branch of research. He is presently employed as Hauptsturmführer (captain) and camp physician in the Auschwitz concentration camp. Anthropological testing of the most diverse racial groups in this concentration camp is being carried out with permission of the SS Reichsführer (Himmler)."[40]

Never one to place principle before pragmatism, the Rockefeller Foundation ceased its funding of most Nazi eugenics, when the Nazis invaded Poland in 1939. By that time, what had been established with their money over a period of more than 15 years was consolidated. Alan Gregg, the Foundation's Director of the Medical Division, was the man who was most intimately involved in the Nazi funding of eugenics at every step of the way. His division was responsible for funding the various Kaiser Wilhelm Institutes.

Another pivotal figure was Raymond B. Fosdick, who became President of the Rockefeller Foundation in 1936, and was by informed accounts, the leading figure in the American Eugenics Society. Fosdick had earlier been general counsel to Sanger's

American Birth Control League, and was the person who in 1924 first convinced John D. Rockefeller Jr. of the importance of birth control and eugenics. He was the brother of prominent eugenics advocate, Harry Emerson Fosdick, the Rockefellers' pastor for whom Rockefeller built the Riverside Church in the mid-1920's. Raymond Fosdick had worked for the Rockefeller family since 1913. He had been sent to the Paris Peace Conference in 1919 as part of Col. Edward Mandell House's group, "The Inquiry," the secret team which ran the American negotiators at Versailles. After Versailles Fosdick went on to become John D. Rockefeller's personal attorney and ran the Rockefeller Foundation for over three decades.[41] In 1924, Fosdick had written a personal letter to John D. Rockefeller Jr., urging foundation funding for Margaret Sanger's eugenics work in birth control, stating: "I believe that the problem of population constitutes one of the great perils of the future and if something is not done along the lines these people are suggesting, we shall hand down to our children a world in which the scramble for food and the means of subsistence will be far more bitter than anything at present."[42]

Leaving Mengele holding the proverbial bag, Verschuer fled Berlin before the end of the war, and avoided a Nuremburg trial. By 1946, he was writing to his old friend, the US Army eugenicist, Paul Popenoe, in California, who had mailed cocoa and coffee to Verschuer back in postwar Germany. Old Nazi friends managed to whitewash Verschuer's Auschwitz past, for which all records had been conveniently destroyed.

In 1949, the Auschwitz doctor Otmar Freiherr von Verschuer was named Corresponding Member of the American Society of Human Genetics, a new organization founded in 1948 by leading eugenicists hiding under the banner of the less-disgraced name, genetics. The first president of the American Society of Human Genetics was Hermann Josef Muller, a Rockefeller University Fellow who had worked at the Kaiser Wilhelm Institute for Brain Research in 1932.[43]

Von Verschuer got his membership in the American Society of Human Genetics from another German, an old eugenics colleague, Dr. Franz J. Kallmann, who had worked with Ernst Rüdin on "genetic psychiatry." Part of von Verschuer's re-packaged identity was a position he got after the war with the newly-created Bureau of Human Heredity in Copenhagen. The Rockefeller Foundation provided the money to found the new Danish office where the same eugenics activities could be more quietly advanced. The Bureau of Human Heredity received a letter from von Verschuer mentioning that he had the results of Auschwitz "research" moved to Copenhagen in 1947, to the care of Danish Institute Director, Tage Kemp, also a member of the American Eugenics Society. Kemp had worked on eugenics with the Rockefeller Foundation since they financed his 1932 research stay at Cold Springs Harbor Eugenics Record Office. Kemp's Institute hosted the first International Congress in Human Genetics after the war in 1956.[44]

JDR III's Population Council and the "Crypto-eugenics"
Eugenics was the foundation of John D. Rockefeller III's obsession with overpopulation. Given his enormous influence and the huge financial muscle of the Rockefeller Foundation to fund scientific research, it was an obsession which would have untold consequences for generations after his death.

John D. III was nurtured on the grim pseudo-science of Malthus and fears of population growth. When he was a senior at Princeton University in 1928, his father, John D. Rockefeller Jr. named his son to the board of the family's Bureau of Social Hygiene, a birth control organization. JDR III's Princeton mentor, economics professor, Frank Fetter, was a member of the American Eugenics Society. Fetter taught that "democracy was increasing the mediocre and reducing the excellent strains of stock...."[45]

In 1931, JDR III joined the board of the Rockefeller Foundation itself. There, eugenicists like Raymond Fosdick and Frederick Osborn, founding members of the American Eugenics Society, fostered JDR III's interest in population control. Osborn was president of the American Eugenics Society in 1946, and was also president of the racist Pioneer Fund. With John D Rockefeller III he would co-found the Rockefeller Population Council. During the Third Reich, Osborn had expressed his early support for German sterilization efforts. In 1937, Frederick Osborn personally praised the Nazi eugenic program as the "most important experiment which has ever been tried."[46] In 1938, he lamented the fact that the public opposed "the excellent sterilization program in Germany because of its Nazi origin." By 1934, a year after Hitler came to power in Germany, JDR III had written his father that he wanted to devote his energies to the population problem.[47]

In 1952, John D. Rockefeller III was ready to begin his life's major work. With $1,400,000 of his own funds in addition to Rockefeller Foundation money, he founded the Population Council in New York, to promote studies on the dangers of "over-population" and related issues. Many of the leading American eugenicists had become pessimistic that their decades of efforts of forced sterilization of mental and other deficient people were making a difference in the quality of the leading genetic stock. With population control, Rockefeller and others in the establishment believed they had finally found the answer to mass, efficient and effective negative eugenics.

John Foster Dulles, then chairman of the Rockefeller Foundation and later Dwight Eisenhower's Secretary of State, played a key role in establishing John D. III's new Population Council along with Frederick Osborn, first director of the Council. Osborn remained a central figure at the Population Council until the late 1960's.

The founding meeting of the Population Council, held in the Rockefeller family's Williamsburg Virginia village, was attended as well by Detlev W. Bronk, then president of both the Rockefeller Institute and the National Academy of Sciences. John D. Rockefeller III arranged for the conference to be sponsored under the auspices of the National Academy of Sciences to give it a quasi-scientific aura. The head of the Academy, Dr. Detlev Bronk, was sympathetic to the agenda of population control. Being promoted was the same unvarnished eugenics racial ideology, veiled under the guise of world hunger and population problems.

In addition, a representative from the Carnegie Institute, Warren S. Thompson, director of the Scripps Foundation for Research in Population Problems, and Thomas Parran, US Surgeon General during the infamous Tuskegee syphilis study, were there as well.

Pascal K. Whelpton from the Population Division at the United Nations came, and so did two men who ran the UN Population Division in later years, Frank Notestein and Kingsley Davis, both as well, members of the American Eugenics Society.[48]

Over the following 25 years, the Rockefeller Population Council would spend a staggering $173 million on population reduction globally, establishing itself as by far the most influential organization promoting the eugenics agenda in the world. Among the Council's favourite projects were funding research for Norplant, a contraceptive steroid implanted under the skin to provide contraception for several years, the IUD contraceptive device, and the French abortion pill, RU-486. Sheldon J. Segal, led the work.[49]

In 1952, when he decided to create the Population Council, Rockefeller scrupulously avoided using the term "eugenics." Population control and family planning were to be the new terms of reference for what was the same policy after 1952, deploying vastly enlarged international resources. The old talk of racial purity and elimination of inferiors was altered. The eugenics leopard, however, did not change its spots after the war. It became far more deadly under John D.'s Population Council. At the time of the founding of Rockefeller's Population Council, the American Eugenics Society made a little-publicized move of its headquarters from Yale University directly into the offices of the Population Council in the Rockefeller Center in New York City.

Rockefeller shrewdly repackaged his discredited eugenics race and class ideology as "population control." Instead of focussing on domestic issues such as American poor immigrants or the mentally challenged, he turned his sights on the entire developing world, a vast sea of humanity which stood between the Rockefeller family and the realization of their ambitious postwar designs for a new American Century.

The strategists around the Rockefeller eugenics organizations explicitly set out to pursue the same agenda as essentially Von Verschuer and the Nazi eugenics crowd had, but under the deliberate strategy of what they termed "crypto-genetics." The key American proponent of hiding the eugenics nature of their work under the name "genetics" and "population control" was Rockefeller's head of the Population Council, Frederick Osborn. Osborn pointed to studies indicating that, with the proper approach, "less intelligent" women can be convinced to reduce their births voluntarily. "A reduction of births at this level would be an important contribution to reducing the frequency of genes which make for mental defect." He asserted that birth control for the poor would help improve the population "biologically." And for families which experience chronic unemployment, Osborn said "Such couples should not be denied the opportunity to use new methods of contraception that are available to better-off families. A reduction in the number of their unwanted children would further both the social and biological improvement of the population." Referring to racial minorities, he explicitly called for "making available the new forms of contraception to the great number of people at the lower economic and educational levels."

"The most urgent eugenic policy at this time," Osborn insisted, was "to see that birth control is made equally available to all individuals in every class of society, because there is new evidence that the more successful or high IQ individuals within each group may soon be having more children than the less intelligent individuals within the group ... these trends are favorable to genetic improvement." He stressed that the reason for making birth control "equally available" should be disguised: "Measures for improving the hereditary base of intelligence and character are most likely to be attained under a name other than eugenics.... Eugenic goals are most likely to be attained under a name other than eugenics."[50]

During McCarthy's Red Scare campaigns in the 1950's United States, countless innocent intellectuals had their careers ruined by being publicly accused of being "crypto-communists," a term denoting one who deeply hides his communist beliefs while working to subvert the American system. In the late 1950's, Dr. Carlos P. Blacker, a past Chairman of the English Eugenics Society, proposed that, "The Society should pursue eugenic ends by less obvious means, that is, by a policy of crypto-eugenics, which was apparently proving successful in the US Eugenics Society."[51] Blacker was close friends with the Population Council's Frederick Osborn.

In 1960, the English Eugenics Society agreed to Blacker's proposal, and adopted a resolution stating, "The Society's activities in crypto-eugenics should be pursued vigorously, and specifically that the Society should increase its monetary support of the Family Planning Association (the English branch of Sanger's Planned Parenthood) and the International Planned Parenthood Federation, and should make contact with the Society for the Study of Human Biology...."[52]

The architect of the American reshaping of the elitist eugenics agenda into the new garment of population control was Rockefeller's friend and employee, Frederick Osborn, first President of John D. Rockefeller III's Population Council, and a founding member of the American Eugenics Society, who was its President until he took the post as head of the Population Council in 1952.

A significant problem after the Second World War was that the very name eugenics had been thoroughly associated in the public mind with Nazi racist extermination programs, the definition of a Master Race and other human atrocities. As Osborn formulated the problem in a 1956 article in *Eugenics Review,* "The very word eugenics is in disrepute in some quarters.... We must ask ourselves, what have we done wrong? We have all but killed the eugenic movement."[53]

Osborn had a ready answer: people for some reason refused to accept that they were "second rate" compared to Osborn, Rockefeller, Sanger and their "superior class." As Osborn put it, "We have failed to take into account a trait which is almost universal and is very deep in human nature. People are simply not willing to accept the idea that the genetic base on which their character was formed is inferior and should not be repeated in the next generation.... They won't accept the idea that they are in general second rate...."[54]

Osborn proposed a change in packaging. Eugenics was to be mass-marketed under a new guise. Instead of talking about eliminating "inferior" people through forced sterilization or birth control, the word would be "free choice" of family size and quality. As early as 1952 when he joined with John D Rockefeller III in the Population Council, Osborn saw the huge potential of contraception and mass education for eugenics, albeit masquerading as free choice. One of his first projects was contributing the funds of his Population Council to research in a new "contraception pill."[55]

"Foreshadowing the future work of the Population Council and the Rockefeller Foundation in population control," Osborn wrote, again in his *Eugenics Review*, "there is certainly a possibility that ... pressures can be given a better direction (for birth control) and can be brought to bear on a majority of the population instead of a minority." And when such pressures are brought to bear, Osborn added, individuals will believe they are choosing on their own not to have children, "if family planning has spread to all members of the population and means of effective contraception are readily available."[56] He wrote that some 13 years before the widespread introduction of the oral birth-control pill.

Osborn went on to call for a system of what he termed, "unconscious voluntary selection." Ordinary people would be led down the path of eugenics and race culling without even being aware where they were going or what they were doing. Osborn argued that the way to convince people to exercise the "voluntary" choice, would be to appeal to the idea of "wanted children." He said, "Let's base our proposals on the desirability of having children born in homes where they will get affectionate and responsible care." In this way, he argued, the eugenics movement "will move at last towards the high goal which Galton set for it," namely creation of the master race, and reduction of inferior races.[57]

Publicly, Osborn appeared to purge eugenics in the postwar era of earlier racism. In reality he applied the racism far more efficiently to hundreds of millions of darker-skinned citizens of the Third World. Osborn also secretly held the office of President of the infamous white-supremacist Pioneer Fund from 1947 to 1956. Among other projects the Pioneer Fund "supported highly controversial research by a dozen scientists who believed that Blacks are genetically less intelligent than Whites," according to a December 11, 1977 article in the *New York Times*.[58] Among recipients of Pioneer Fund money was Stanford University Nobel laureate, William Shockley, who advocated forced sterilization of all persons with an IQ below 100. He got more than $1 million in research funds from Osborn's Pioneer Fund.[59]

When Osborn wrote those words advocating "unconscious voluntary selection," he was still Secretary of the American Eugenics Society and President of John D. Rockefeller III's newly-founded Population Council. JDR III was chairman and Princeton eugenicist, Frank Notestein, was board member, and later became President of Rockefeller's Council.

Yes, Hello Dolly…

Member of the Rockefeller Foundation board and close family friend, Frederick Osborn was an unfettered enthusiast of the Rockefeller Foundation's support for Nazi eugenics experiments. Scion of a wealthy American railroad family, graduated in 1910 from Princeton University, which would later be the school of John D. III, Osborn was a part of the wealthy American upper class. Under the banner of philanthropy, Osborn would pursue policies designed to preserve the hegemony and control of society by his wealthy associates.

In 1937, Osborn had praised the Nazi eugenics program as the "most important experiment that has ever been tried."[60] One year later, Osborn bemoaned the fact that the general public seemed opposed to "the excellent sterilization program in Germany because of its Nazi origin."[61] Osborn and the Rockefeller Foundation knew well that their money was going toward the Third Reich, even though they later piously disavowed this knowledge.

As late as 1946, after the war and the ghastly revelations of the human experiments in Auschwitz and other concentration camps, Osborn, then President of the American Eugenics Society, published, in his *Eugenics News* magazine, the so-called Geneticists' Manifesto entitled, "Genetically Improving the World Population."

In 1968, Osborn published his book, *The Future of Human Heredity: An Introduction to Eugenics in Modern Society*. He had lost his postwar inhibitions about calling his work what it was: eugenics. At that time his nominal boss and protégé in population eugenics, Population Council chairman John D. Rockefeller III, was preparing himself to head a Presidential commission on the population problem.

In his book, Osborn cited studies showing that less intelligent women could be persuaded to reduce their births voluntarily: "A reduction of births at this level would be an important contribution to reducing the frequency of genes which make for mental defect." Osborn added: "The most urgent eugenic policy at this time is to see that birth control is made equally available to all individuals in every class of society." He also noted that: "Eugenic goals are most likely to be attained under a name other than eugenics."[62] In short, they would be most likely attained by using crypto-eugenics tactics. In a speech to the annual meeting of the American Eugenics Society in 1959, Osborn stated, "With the close of World War II, genetics had made great advances and a real science of human genetics was coming into being.... Eugenics is at last taking a practical and effective form."[63] Genetics was the new name for eugenics.

Forerunning the later human cloning debate and the widely publicized sheep clone, Dolly, Osborn doled out strong praise for Hermann J. Muller, Ernst Rüdin's colleague in Germany, who had received Rockefeller funds during the 1930's for eugenics research. Quoting Muller, Osborn wrote, "It would in the end be far easier and more sensible to manufacture a complete new man de novo, out of appropriately chosen raw materials, than to try to refashion into human form those pitiful relics which remained."[64] Osborn also praised Muller's proposal to develop sperm banks to "make available the sperm of highly qualified donors." The idea for a gene revolution was being debated back then.

Rockefeller's Population Council gave grants to leading universities including Princeton's Office of Population headed by Rockefeller eugenicist, Frank Notestein, a long-time friend of Osborn, who in 1959 became President of the Rockefeller Population Council in order to promote a science called demography. Its task was to project horrifying statistics of a world overrun with darker-skinned peoples, to prepare the ground for acceptance of international birth control programs. The Ford Foundation soon joined in funding the various Population Council studies, lending them an aura of academic respectability and above all, money. The Rockefeller Population Council grants were precisely targeted at creating a new cultural view about growing human population through their funding of demographic research such as that of Princeton's Notestein. According to John Sharpless who studied the history of population control using the Rockefeller Foundation archives, in the 1950's:

> the non-profit sector was where the debate over the population problem actually played itself out, ultimately defining how the policy issue would be viewed in the period which followed [emphasis added]. ... [The Population Council made sure that] research would take place in both the social as well as the biological sciences ... this effort was not simply an exercise in pure science but one which aimed specifically at policy ... not only the legitimating of the "science" of demography but also the acceptance of demography as a policy science ... they were slowly encouraging an evolution in thinking among "population specialists" to view intervention in demographic processes (particularly fertility) as not only appropriate but necessary.[65]

In 1952, the same year that John D. founded the Population Council with Osborn at its head, Margaret Sanger created, thanks to Rockefeller Foundation money, a global version of her American Planned Parenthood Federation called the International Planned Parenthood Federation (IPPF). Sanger had first met JDR III in 1947. She had convinced him then of the urgency of promoting mass birth control.

Following the initial Rockefeller financing, her IPPF soon was backed by a corporate board which included DuPont, US Sugar, David Rockefeller's Chase Manhattan Bank, Newmont Mining Co., International Nickel, RCA, Gulf Oil and other prominent corporate members. The cream of America's corporate and banking elite were quietly lining up behind Rockefeller's vision of population control on a global scale.

Less than a decade after the revelations of eugenics and Auschwitz, population control was again becoming fashionable in certain American elite circles of the 1950's. It was a testimony to the power of the US establishment to mould public opinion and encourage fears of exploding populations of poor, hungry peasants around the world.

In 1960, Rockefeller friend and wealthy patron of population control, Hugh Moore, founded the World Population Emergency Campaign with the help of funds from DuPont, which would later become a major promoter of the gene revolution in agriculture. Eugene R. Black, former senior executive of David Rockefeller's Chase National Bank, ran, as president of the World Bank, a campaign which had as its main aim to create and reinforce First World fears of a population explosion in Third World countries.

The 1958 Castro revolution in Cuba provided additional impetus to instigating these fears among unwitting Americans. The argument promoted in the American mass media by circles around the Population Council was simple and effective: over-population in poor developing countries leads to hunger and more poverty, which is the fertile breeding ground for communist revolution.

John D. III's brother, Laurance Rockefeller, established and ran the Conservation Foundation in 1958 to complement John D.'s Population Council. Both the Population Council and Conservation Foundation were united around the unspoken theme that natural resources must be conserved, but conserved from use by smaller businesses or individuals, in order that select global corporations should be able to claim them, thus establishing a kind of strategic denial policy masquerading as conservation.

The population control lobby which would later shape Kissinger's NSSM 200 was consolidating around Rockefeller Foundation grants and individuals, preparing a global assault on "inferior peoples," under the name of choice, of family planning and of averting the danger of "over-population"—a myth their think-tanks and publicity machines produced to convince ordinary citizens of the urgency of their goals.

From Eugenics to Genetics

A colleague of Ernst Rüdin, Dr. Franz J. Kallmann was a German scientist who left Germany in 1936 when it was discovered that he was part Jewish. After the war, he helped rehabilitate German eugenicist Otmar Freiherr von Verschuer and win him respectability and acceptance in the American scientific community. Kallmann's enthusiasm for eugenics was in no way dampened by his own experience of Nazi persecution of the Jews. In addition to teaching at Columbia University, Kallmann was a psychiatric geneticist at the New York State Psychiatric Institute, and in 1948, he was founding President of a new eugenics front organization, the American Society of Human Genetics. At the New York Psychiatric Institute, Kallmann continued the same research in genetic psychiatry he had done with Rüdin in Germany.

Kallmann was a thorough-going advocate of practicing elimination or forced sterilization on schizophrenics. In 1938, when in the United States, he wrote in an article translated by Frederick Osborn's *Eugenics News*, that schizophrenics were a "source of maladjusted crooks, asocial eccentrics and the lowest type of criminal offenders." He demanded the forced sterilization of even healthy offspring of schizophrenic parents to kill the genetic line.[66]

The choice of the term Human Genetics reflected the attempt to disguise the eugenic agenda of the new organization. Most of its founding members were simultaneously members of Frederick Osborn's American Eugenics Society. By 1954, his old friend von Verschuer was also a member of this one big happy eugenics family. Kallmann's American Society of Human Genetics soon got control of the field of medical eugenics, recognized by the American Medical Association as a legitimate medical field.

Kallmann's American Society of Human Genetics later became a sponsor of the Human Genome Project. The multibillion dollar project was, appropriately enough, housed at the same Cold Spring Harbor center that Rockefeller, Harriman and Carnegie had used for their notorious Eugenics Research Office in the 1920's. Genetics, as defined by the Rockefeller Foundation, would constitute the new face of eugenics.

While brother John D. III was mapping plans for global depopulation, brothers Nelson and David were busy with the business side of securing the American Century for the decades following the crisis of the 1960's and 1970's. American agribusiness was to play a decisive role in this project, and the development of genetic biotechnology would bring the different efforts of the family into a coherent plan for global food control in ways simply unimaginable to most.

Notes

1. Carroll Quigley, *Tragedy and Hope: A History of the World in Our Time*, New York, The Macmillan Co., 1966, p. 842. Quigley details the transfer of Operations Management techniques from the military after World War II to Operation Bootstrap under the Puerto Rican Industrial Development Corporation run by Governor Munoz via the US pentagon consultancy, Arthur D. Little Inc. Laurence Rockefeller used Bootstrap state funds to build a luxury Dorado Beach Hotel and Golf Club (see The Rockefeller Archive Center, in http://archive.rockefeller.edu/bio/laurance.php#lsr6). North American Congress on Latin America (NACLA), *Puerto Rico to New York: the Profit Shuttle*, April 1976, NACLA Digital Archive, provides details about the role of Rockefeller's Chase Bank and Nelson Rockefeller's IBEC in Operation Bootstrap.

2. NACLA, *op. cit.,* pp. 11-12.

3. "In 1950 and 1951, John Foster Dulles, then chairman of the Rockefeller Foundation, led John D. Rockefeller III on a series of world tours, focusing on the need to stop the expansion of the non-white populations. In November 1952, Dulles and Rockefeller set up the Population Council, with tens of millions of dollars from the Rockefeller family," cited in *Eugenics, a brief history*, in http://www.tribalmessenger.org/t-secret-gov/eugenics.htm.

4. Dr. J. L. Vasquez Calzada study cited in Bonnie Mass, "Puerto Rico: A Case Study of Population Control," *Latin American Perspectives,* Vol. 4, No. 4, Fall 1977, pp. 66-81.

5. Charles W. Warren, et al., "Contraceptive Sterilization in Puerto Rico," *Demography*, Vol. 23, No. 3 (Aug., 1986), pp. 351-352.

6. Susan E. Lederer, *"Porto Ricochet": Joking about Germs, Cancer, and Race Extermination in the 1930s*, Oxford University Press, Oxford, 2002, p. 732. Also "Porto Ricochet," *Time*, 15 February 1932, in http://www.time.com/time/magazine/article/0,9171,743163,00.html for the quote by Rhoads.

7. Time, *op.cit.*; see also, Douglas Starr, "Revisiting a 1930s Scandal: AACR to Rename a Prize," *Science*, vol. 300, no. 5619, 25 April 2003, pp. 573-574.

8. *Ibid.* The ACHE Report on the Rhoads cancer experiments is available at http://www.seas.gwu.edu/nsarchive/radiation/. See also Stycos, J.M., "Female Sterilization in Puerto Rico", *Eugenics Quarterly*, no.1, 1954.

9. John D. Rockefeller III, *People, Food and the Well-Being of Mankind*, Second McDougall Lecture, Food and Agriculture Organization of the United Nations, 1961, pp. 9, 16-18.

10. Jameson Taylor, *Robbing the Cradle: The Rockefellers' Support of Planned Parenthood*, http://www.lifeissues.net/writers/tay/tay_04robthecrad.html.

11. Alan Gregg cited in Julian L. Simon, "*The Ultimate Resource II: People, Materials, and Environment,*" chapter 24: Do Humans Breed Like Flies?, Princeton University Press, Princeton, 1996, p.343-4.

12. Alan Gregg, "A Medical Aspect of the Population Problem," *Science,* vol. 121, no. 3150, 13 May 1955, pp. 681-682.

13. Andrew Carnegie, "Wealth", *North American Review,* June 1889, p. 653.

14. John Ensor Harr and Peter J. Johnson, *The Rockefeller Century: Three Generations of America's Greatest Family,* Scribner's, New York, 1988, pp. 452-453.

15. Margaret Sanger, "The Eugenic Value of Birth Control Propaganda", *Birth Control Review,* October 1921, p. 5.

16. Margaret Sanger was clear in her advocacy of racial superiority. In 1939 she created The Negro Project. In a letter to a friend about the project, she confided, "The minister's work is also important and he should be trained, perhaps by the Federation as to our ideals and the goal that we hope to reach. *We do not want word to go out that we want to exterminate the Negro population, and the minister is the man who can straighten out that idea if it ever occurs to any of their more rebellious members*" [emphasis added] cited in Tanya L. Green, *The Negro Project: Margaret Sanger's Genocide Project for Black Americans,* http://www.blackgenocide.org/negro.html. The board of Sanger's Planned Parenthood Federation, which received generous funding from the Rockefeller Foundation, included some of the most prominent eugenicists of the day. Lothrop Stoddard, a Harvard graduate and the author of *The Rising Tide of Color against White Supremacy,* was a Nazi enthusiast who described the eugenic practices of the Third Reich as "scientific" and "humanitarian". Dr. Harry Laughlin, another Sanger board member, spoke of purifying America's human "breeding stock" and purging America's "bad strains" which he defined to include, "the shiftless, ignorant, and worthless class of anti-social whites of the South". Laughlin was the Superintendent of the Eugenics Record Office from 1910 to 1921; he was later President of the Pioneer Fund, a white supremacist organization that is still functioning today.

17. Margaret Sanger, *The Pivot of Civilization,* Brentano's Press, New York, 1922, p. 189.

18. Cold Spring Harbor Laboratory, *Archives: Eugenics Record Office,* in http://library.cshl.edu/archives/archives/eugrec.htm.

19. Harry Laughlin, *Report of the Committee to Study and to Report on the Best Practical Means of Cutting Off the Defective Germ-Plasm in the American Population,* Cold Spring Harbor, New York, 1914, p. 1. The project was a joint undertaking by the American Breeders' Association and the Cold Spring Harbor Eugenics Record Office.

20. Edwin Black, *War Against the Weak: Eugenics and America's Campaign to Create a Master Race,* Thunders' Mouth Press, New York, 2004, p. 57. See also "Extends Work In Eugenics, Harriman Philanthropy to Have a Board of Scientific Directors", *The New York Times,* 20 March 1913, which cites Rockefeller financial support to the Eugenics Records Office in 1913 as second only to Mrs. E. Harriman.

21. Paul Popenoe and R. H. Johnson,, *Applied Eugenics,* Macmillan Company, New York, revised edition, 1933, p. 135.

22. *Ibid.* pp. 123-137.

23. Oliver Wendell Holmes, *Carrie Buck vs. J.H. Bell,* The Supreme Court of the United States, No. 292, October Term, 1926, p. 3.

24. The justices of the Supreme Court never met Buck. They relied on the expert opinion of Dr. Harry Hamilton Laughlin, head of Eugenics Record Office, in Cold Spring Harbor, New York to help them make up their minds. Though Laughlin had never met her either, a report had been sent to him, including Buck's score on the Stanford-Binet test that purportedly showed Buck had the intellectual capacities of a nine-year-old. Laughlin concluded that she was part of the "shiftless, ignorant and worthless class of anti-social whites of the South' whose promiscuity offered "a typical picture of the low-grade moron." Laughlin quotes are cited in Peter Quinn, "Race Cleansing in America", *American Heritage Magazine,* February/March 2003. The 1922 quote from Justice Holmes is contained in a letter from Oliver Wendell Holmes Jr., to Harold J. Laski, 14 June 1922, *Holmes-Laski Letters Abridged,* edited by Mark DeWolfe Howe Atheneum, Clinton, MA, 1963, Vol. 1, p. 330.

25. State of Illinois Board of Administration, *Vol. II: Bienniel Reports of the State Charitable Institutions: October 1, 1914 to September 30, 1916,* State of Illinois, 1917, p. 695, cited in Edwin Black, *op. cit.,* pp. 254-255.

26. Edwin Black, *op. cit.,* p. 122.

27. Paul Weindling, "The Rockefeller Foundation and German Biomedical Science, 1920-40: from Educational Philanthropy to International Science Policy", in N. Rupke (editor), *Science, Politics and the Public Good. Essays in Honour of Margaret Gowing,* Macmillan, Basingstoke, 1988, pp. 119-140. Reprinted: G. Gemelli, J-F Picard, W.H. Schneider, *Managing Medical Research in Europe. The Role of the Rockefeller Foundation (1920s-1950s),* CLUEB, Bologna, 1999, p. 117-136. See also Stefan Kuehl, *The Nazi Connection: Eugenics, American Racism, and German National Socialism,* Oxford University Press, Oxford, 1994, pp. 20-21.

28. Rockefeller Foundation Archives, Series 717 A: Germany, Box 10, Folder 64, *Kaiser Wilhelm Institute, Berlin—Brain Research,* 1928-1939, in http://archive.rockefeller.edu/publications/guides/psychiatry.pdf.

29. Dr. Julius Hallervorden, Institute for Brain Research, in testimony to his interrogators after the war, quoted in Michael Shevell, "Racial Hygeine, Active Euthanasia and Julius Hallervorden", *Neurology*, volume 42, November 1992, pp. 2216-2217.

30. Ernst Rüdin, "Hereditary Transmission of Mental Diseases", *Eugenical News*, Vol. 15, 1930, pp. 171-174. Also, D.P.O'Brien, *Memorandum from D.P. O'Brien to Alan Gregg*, 10 November 1933, Rockefeller Foundation, RF 1.1 717 946. Cited in Edwin Black, *op. cit.*, p. 296. See as well, Cornelius Borck, *The Rockefeller Foundation's Funding for Brain Research in Germany, 1930-1950*, Rockefeller Center Archive Newsletter Spring 2001, http://www.archive.rockefeller.edu/publications/newsletter/nl2001.pdf. Borck, a German researcher, was given permission to visit the Rockefeller Center archives to study files relating to the foundation's support for brain research during the Third Reich and after. Though his report is very mild, he is forced to admit a number of embarrassing items: "the RF (Rockefeller Foundation-ed.) did not cease its activities in Germany in 1933; indeed, it did not do so until the United States entered into World War II." And further: "the RF had funded, during the 1920s and early 1930s, some projects by individual scientists engaged in eugenics and hereditary diseases who soon became close allies of the new regime and its ambitions for a racial science, such as, for example, Ernst Rüdin's program of an epidemiology of inherited nervous and psychiatric disease, or Walther Jaensch's outpatients' clinic for constitutional medicine at the Charité."

31. "Eugenical Sterilization in Germany", *Eugenical News*, Vol. 18, 1933, pp.91-93.

32. Edwin Black, *op. cit.*, p. 299.

33. Thomas Ruder and Volker Kubillus, "Manner Hinter Hitler", *Verlag fur Politik und Gessellshaft*, Malters, 1994, pp. 65-66.

34. Adolf Hitler, *Mein Kampf*, translated by Alvin Johnson, Reynal & Hitchcock, New York, 1941, Vol. 2, Chapter 3, p. 658.

35. Madison Grant, *The Passing of the Great Race*, Charles Scribner's Sons, New York, 1936, pp.50 51, 89.

36. Leon Whitney quoted in Edwin Black, *op. cit.*, p. 317.

37. Radiogram to Alan Gregg, 13 May 1932, Rockefeller Foundation RF 1.1 Ser 7171 Box 10 Folder 63, cited in Edwin Black, *op. cit.*, p. 297.

38. Raymond B. Fosdick, Letter to Selskar M. Gunn, 6 June 1939, Rockefeller Foundation RF 1.1 717 16 150, cited in Edwin Black, *op. cit.*, p 365. According to Gunn, the Foundation's official denials of funding Nazi research were "of course hardly correct."

39. Edwin Black, *op. cit.*, p. 341.

40. Otmar Freiherr von Verschuer cited in Edwin Black, "Eugenics and the Nazis—the California connection", *San Francisco Chronicle*, 9 November, 2003.

41. Eugenics Watch, *Eugenics: An Antidemocratic Policy*, http://www.eugenics-watch.com/eugbook/euod_ch1.html#6.

42. Raymond D. Fosdick to John D. Rockefeller Jr., cited in Rebecca Messall, *The Long Road of Eugenics: from Rockefeller to Roe v. Wade*. Originally published in *Human Life Review*, Vol. 30, No. 4, Fall 2004, pp. 33-74, http://orthodoxytoday.org/articles5/MessallEugenics.php.

43. Edwin Black, *op. cit.*, p. 379.

44. Tage Kemp, *Report of Tage Kemp to the Rockefeller Foundation*, 17 November 1932, RF RG 1.2, Ser 713, Box 2, Folder 15, cited in Edwin Black, *op. cit.*, pp. 418-419. Also Benno Müller-Hill, *Die ödliche Wissenschaft: Die Aussonderung von uden, Zigeunern und Geisteskranken 1933-1945*, Rowohlt, Reinbeck bei Hamburg, 1984, p. 129.

45. Thomas C. Leonard, "Retrospectives: Eugenics and Economics in the Progressive Era", *Journal of Economic Perspectives*, Fall 2005, p. 210. John Ensor Harr and Peter J. Johnson, *The Rockefeller Century: Three Generations of America's Greatest Family*, Scribner's, New York, 1988, p. 272.

46. Frederick Osborn, *Summary of the Proceedings of the Conference on Eugenics in Relation to Nursing, 24 February, 1937*, American Eugenics Society Papers: Conference on Eugenics in Relation to Nursing, cited in Stefan Kühl, *op.cit.*, pp. 40-41.

47. *Ibid.* On Fosdick's influence in shaping John D. Rockefeller III's interest in eugenics and population see Harr & Johnson, *op. cit.*, p. 369.

48. John Cavanaugh-O'Keefe, *The Roots of Racism and Abortion: An Exploration of Eugenics, Chapter 10: Eugenics after World War II*, 2000, http://www.eugenics-watch.com/roots/index.html.

49. Population Council, "The ICCR at 30: Pursuing New Contraceptive Leads", *Momentum: News from the Population Council*, July 2000.

50. Frederick Osborn, *The Future of Human Heredity: An Introduction to Eugenics in Modern Society*, Weybright and Talley, New York, 1968, pp. 93-104. Curiously, Osborn never dropped his use of the term eugenics even in 1968.

51. John Cavanaugh-O'Keefe, *op. cit.*, Chapter 10: Eugenics after World War II, C. P. Blacker and "Crypto-Eugenics."

52. *Ibid.*

53. *Ibid.*

54. *Ibid.*

55. *Ibid.*

56. *Ibid.*

57. *Ibid.*

58. Grace Lichtenstein, "Fund Backs Controversial Study of Racial Betterment", *The New York Times*, 11 December 1977. The article states, "A private trust fund based in New York has for more than 20 years supported highly controversial research by a dozen scientists who believe that blacks are genetically less intelligent than whites... A month-long study of the Pioneer Fund's activities by The New York Times shows it has given at least $179,000 over the last 10 years to Dr. William B. Shockley, a leading proponent of the theory that whites are inherently more intelligent than blacks."

59. Lichtenstein, *op. cit.*

60. Stefan Kühl, *op.cit.*, pp. 40-41

61. *Ibid.*

62. Frederick Osborn, *op. cit.*, pp. 93-104.

63. Frederick Osborn, *Eugenics: Retrospect and Prospect, Draft Prepared for the Directors' Meeting, April 23rd*, Draft of 26 March 1959, American Philosophical Society, AES Records—Osborn Papers, cited in Edwin Black, *War Against the Weak*, p. 423.

64. Frederick Osborn, *The Future of Human Heredity: An Introduction to Eugenics in Modern Society*, Weybright and Talley, New York, 1968, pp. 93-104.

65. John B. Sharpless, *The Rockefeller Foundation, the Population Council and the Groundwork for New Population Policies*, Rockefeller Archive Center Newsletter, Fall 1993.

66. John Cavanaugh-O'Keefe, *op. cit.*, Chapter 10, *The Shift to Genetics*.

CHAPTER 6
Fateful War and Peace Studies

Preparing a Post-War Empire

W ell before the triumphant victory of the United States in World War II, it had become obvious to the heads of the largest American corporations and banks that the US market was far too small for their ambitions. As they saw it, "Manifest Destiny," the unlimited expansion of American power, was to be a worldwide business. A seemingly easy victory in World War I and the gains of the Versailles Treaty in Europe had only whetted their appetite for more.

Leading policy-making figures of the American establishment had quietly created a highly influential policy group in late 1939, only weeks after the German invasion of Poland, and two full years before Pearl Harbor would bring the US directly into the war. The task of the secret group was simple: to shape US post-war economic and political goals, based on the assumption a world war would come and that the United States would emerge from the ashes of that war as the dominant global power.

That elite policy-making circle, the War and Peace Studies Group of the New York Council on Foreign Relations, effectively took over all significant post-war planning for the US State Department. After 1942, most of its members were quietly put directly on the State Department payroll.

Their work was financed by the ubiquitous Rockefeller Foundation. Between November 1939 and late 1942, the Rockefeller Foundation had contributed no less than $350,000 to finance the drafting of the agenda for post-war American economic hegemony via the War & Peace Studies Group. It was an investment which, like most made by the Foundation, paid back thousands-fold in later years. It defined the post-war American business empire globally.[1]

During the interwar years of the 1930's, while most Americans were struggling with the devastation of the Great Depression, a handful of businessmen and their academic associates at private universities such as Harvard, Yale, Princeton, and Johns Hopkins, along with senior partners from the major Wall Street law firms, were preparing the ground for the new "Pax Americana." Their aim was simple: to consolidate an American succession to the failing Pax Britannica of the British Empire.

These American policymakers were largely concentrated among the select membership of the New York Council on Foreign Relations. Unlike the British Empire, their American vision of global domination was based on economic goals rather than physical possession of a colonial empire. It was a brilliant refinement which allowed the US corporate giants to veil their interests behind the flag of democracy and human rights for "oppressed colonial peoples," support of "free enterprise" and "open markets."

The interests represented in the Council on Foreign Relations task force were anything but democratic. It was that of the elite handful of American corporations and their law firms which had developed global interests, namely in oil, banking and related industries. The businessmen represented in the Council on Foreign Relations, or CFR as it was called, were a breed apart. They were no ordinary small entrepreneurs.

The CFR had been established in May 1919, in the days of the Versailles Peace conference in an exclusive meeting at the Paris Hotel Majestic, by leading representatives of the J.P. Morgan bank, including Thomas Lamont, together with representatives of the Rockefellers' Standard Oil group, and other select persons including Woodrow Wilson's adviser, Col. Edward House. They met together with equally select British friends, most members of Cecil Rhodes' secretive Round Table group, to discuss establishing a private network of institutes to "advise" their respective governments on foreign affairs.

The handful of influential US banks and corporations going abroad in the era of World War I were few. Most were headquartered in New York on the East Coast, leading some to refer to it as the East Coast Establishment. Its de facto headquarters after World War I was the newly founded Council on Foreign Relations in New York. The initial financing to establish the CFR came from J.P. Morgan, John D. Rockefeller, financiers Otto Kahn, Bernard Baruch, Jacob Schiff and Paul Warburg, the most powerful men of their day in American business.[2]

This elite group had been successful in opening the legal doors for their move overseas by lobbying for a series of Congressional acts that exempted them from prohibitions against monopoly and other US Government anti-trust restrictions.

In 1918, Congress passed the Webb-Pomerene Act, which exempted companies from anti-trust laws, effectively permitting monopolies, "if their activities are directed to export promotion." Standard Oil was a major beneficiary of this act. In 1919, Congress passed the Edge Act, which exempted US banks from the same anti-trust laws for export activity and export of capital. Chase Bank, National City Bank and J.P. Morgan in New York were the main beneficiaries of the Edge Act. Furthermore, in 1920, the US Supreme Court ruled in the US Steel case that mergers creating a near total market control were "not necessarily against the public interest."[3] At the core of these foreign US interests during the 1920's were the leading banks and oil interests of the Rockefeller and Morgan families.

They were the international corporate industry and banking leaders who had already seen, up close, what lucrative potentials existed in taking over the shards of European colonial empires. Compared with what they saw as the limited market potentials within the bounds of the United States, domination of vast new foreign markets offered untold potential, profits and, above all, power.

The "American Century"—The US Lebensraum

In early 1941, some ten months before the Japanese bombing of Pearl Harbor, Henry Luce, publisher of *Time* and *Life* magazines, and a well-connected member of the East Coast elite, wrote an editorial in the February 17 issue of *Life* entitled, "The American Century." In his essay, Luce described the emerging consensus of the US East Coast establishment around the CFR.

"Tyrannies," Luce wrote, "may require a large amount of living space; but Freedom requires and will require far greater living space than Tyranny." He made an open call for Americans to embrace a new role as the dominant power in the world, a world in which the United States had not yet entered the war. He wrote, "the cure is this: to accept wholeheartedly our duty and our opportunity as the most powerful and vital nation in the world and in consequence to exert upon the world the full impact of our influence, for such purposes as we see fit and by such means as we see fit."[4]

Luce was reflecting the emerging view of the internationally-oriented US business and banking establishment around Morgan and Rockefeller. They needed unfettered access to global resources and markets after the war, and they saw the golden chance to get it while all contending powers had been devastated by war.

The American banking and industrial giants needed room, or what some called a Grand Area. The Economic & Financial Group of the CFR War & Peace Studies made a survey of world trade in the late 1930's. They proposed linking the Western Hemisphere with the Pacific into a US-dominated bloc, which was premised on what they called "military and economic supremacy for the United States."[5] The bloc included what was then, still, the British Empire. Their Grand Area was to encompass most of the planet, outside the sphere of the Soviet Union which, to their irritation, remained closed to American capital penetration.

Founding CFR member and one of the leaders of the CFR War & Peace Study group, Isaiah Bowman, known as "America's Geopolitician" during the Second World War, had another term for the Grand Area. Bowman called it, in reference to Hitler's geographical term for the economic justification of German expansion, "an American economic Lebensraum."[6] The term was later dropped for obvious reasons, and the more neutral-sounding American Century was used instead to describe the emerging vision of post-war US imperialism.

As Bowman and others of the CFR State Department study group saw it, the champions of the new American economic geography would define themselves as the selfless advocates of freedom for colonial peoples and as the enemy of imperialism. They would champion world peace through multinational control. Since late days of World War I, when Bowman had worked on The Inquiry, a top-secret strategy group of President Woodrow Wilson, Bowman had been occupied with how to clothe American imperial ambitions in liberal and benevolent garb.

As Bowman and other CFR planers envisioned it, the American domination of the world after 1945 would be accomplished via a new organization, the United Nations, including the new Bretton Woods institutions of the International Monetary Fund and World Bank, as well as the General Agreement on Tariffs and Trade (GATT).

Bowman's CFR group had drafted the basic outline for President Roosevelt of what would become the United Nations Organization. Under the banner of "free trade" and the opening of closed markets around the world, US big business would advance their agenda, forcing open new untapped markets for cheap raw materials as well as new outlets for selling American manufactures after the war.

The group drafted more than 600 policy papers for the State Department and President Roosevelt, covering every conceivable part of the planet, from Continents to the smallest islands. It was all based on a presumed US victory in a war which Washington was not even officially fighting.

For the CFR and the forward-looking members of the US policy-making establishment, after World War II, global power would no longer be measured in terms of military control over colonial territories. The British and European empires proved to be a system far too costly and inefficient. Power would be defined directly in economic terms. It would be based on what one Harvard proponent, Joseph Nye, later was to call "soft power."[7]

As the War came to an end in 1945, no group epitomized the global outlook of American big business more than the Rockefeller family, whose fortune had been built on a global empire of oil and banking. The family—above all brothers Nelson, John D. III, Laurance and David—whose foundation had financed the War & Peace Studies of the CFR, viewed the victorious end of the War as a golden opportunity to dominate global policies to their advantage as never before.

Nelson Aldrich Rockefeller was to play a discreet and decisive behind-the-scenes role in defining those global interests. They were shrewdly redefined from being Rockefeller private interests, into what was called "American national interests." After all, the family had financed the War & Peace Studies for the State Department.

Nelson Ventures in Latin America
Precisely what Isaiah Bowman and his War and Peace Studies colleagues in the US establishment had in mind with their notion of Grand Area and free market development soon became clear. Nelson Rockefeller, one of the primary financial backers of the Council on Foreign Relations War & Peace Studies, wasted no time in taking advantage of the new economic possibilities World War II had opened up for American business.

After the War, while brother John D. Rockefeller III was busy devising new, ever more efficient methods to promote racial purity and depopulation through his Population Council, Nelson was working the other side of the fence. It was in the role of a forward-looking international businessman interested in making world food production, especially in poorer, less-developed countries such as Mexico, more "efficient." Nelson later called his revolution in world agriculture the Green Revolution. It was revolutionary, but not in the way most people had been led to believe.

During the War itself, Nelson had combined promotion of the vast Rockefeller family interests throughout Latin America, with a senior US Government intelligence position, Co-ordinator of Inter-American Affairs (CIAA), nominally on behalf of the Roosevelt White House. From that strategic position, Nelson could funnel US Government support to Rockefeller family business allies in key countries, from Brazil to Peru, Mexico, Venezuela and even Argentina, under the guise of combating Nazi infiltration of the Americas and of promoting "American democracy." He was carefully laying the basis for post-war American business expansion.[8]

Nelson was named as CIAA head in August 1940, in a clear violation of US official neutrality. To conceal that delicate point, the CIAA was given a cover as an organization promoting "American culture" in Latin America.

Skeletons in Rockefeller's Dark Closet

In 1941, Standard Oil of New Jersey, later renamed Exxon, was the largest oil company in the world. It controlled 84% of the US petroleum market. Its bank was Chase Bank, and its main owners were the Rockefeller group. After the Rockefellers, the next largest stockholder in Standard Oil was I.G. Farben, the enormous petrochemicals trust of Germany, which at the time was a vital part of the German war industry. The Rockefeller—I.G. Farben relationship went back to 1927, around the same time the Rockefeller Foundation began heavily funding German eugenics research.[9]

While Nelson Rockefeller was ostensibly combating Nazi economic interests in Latin America as head of the CIAA, the Rockefeller family's Standard Oil, through its President, Walter Teagle, was arranging to ship vital tetraethyl lead gasoline to the German Luftwaffe. When Britain protested the shipment of such strategic materials to Nazi Germany, as Britain itself was being bombed by German Luftwaffe planes, Standard Oil changed its policy. The change was purely cosmetic. They merely altered the registration of their entire fleet to Panama to avoid British search or seizure. Their ships continued to carry oil to Tenerife in the Canary Islands, off the coast of Morocco and the Spanish Sahara in Northwest Africa, where they refueled and siphoned oil onto German tankers for shipment to Hamburg.[10]

During the war, US Senator Harry S. Truman charged, in a Senate investigation, that the Rockefeller-I.G. Farben relationship "was approaching treason."[11] CBS News war correspondent, Paul Manning, reported that on August 10, 1944, the Rockefeller-I.G. Farben partners moved their "flight capital" through affiliated American, German, French, British and Swiss banks.

Nelson Rockefeller's role in Latin America during the War was to coordinate US intelligence and covert operations in the days before the creation of the CIA. He was the direct liaison between President Franklin Roosevelt and British Prime Minister Winston Churchill's personal intelligence head for the Americas, Sir William Stephenson, who directed a front company called British Security Coordination or BSC. Notably, Stephenson's clandestine headquarters for his covert activity was in room 3603 in Rockefeller Center, in New York City, not far from Nelson's office. It was no coincidence. Rockefeller and Stephenson coordinated closely on mutual intelligence operations in the Americas.[12]

Rockefeller brought with him to Washington a team he selected from family business connections, including Joseph Rovensky from Chase Bank, and Will Clayton, a Texas cotton magnate from the agricultural commodity firm Anderson Clayton.[13] Nelson's assistant, John McClintock, ran the vast United Fruit plantations across Central America after the war, on whose behalf the CIA later conveniently orchestrated a coup in Guatemala in 1954.

During the war, Nelson Rockefeller's work laid the basis for the family's vast expansion of interests in the 1950's. He shaped a US-Latin American defense concept which was to tie the military elite of the region to US policies during the Cold War, often through ruthless military dictators who benefited from the backing of the Rockefeller family and insured favorable treatment of Rockefeller business interests. Nelson called the cooperative Latin military dictators he backed, "the New Military."[14]

Nelson Rockefeller had been a leading figure in US corporate investment in Latin America since the 1930s, a time when he was a director of Standard Oil's Venezuelan subsidiary, Creole Petroleum. In 1938, he had tried, and failed, to negotiate a settlement with Mexico's President Lazaro Cardenas for Standard Oil in Mexico. Cardenas had nationalized Standard Oil, leading to bitter US-Mexico relations.

In the 1940s, Rockefeller set up the Mexican American Development Corp. and was a personal investor in Mexican industries after the war. He encouraged his brother David to set up Chase Bank's Latin American division. One motive was to regain a foothold in Mexico through the guise of helping to solve the country's food problems.[15]

As chairman of the US Government's International Development Advisory Board, Rockefeller became the architect of President Harry S. Truman's foreign-aid program. Typically, Nelson used US guarantees to leverage the massive private lending by Chase, National City Bank (today Citigroup Inc.) and other New York banks throughout the Latin American region.

During the War, as head of Roosevelt's CIAA, Nelson had organized a network of journalists and of major newspapers owners throughout the region. He did this by threatening neutral Latin American newspaper publishers with a cut-off of newsprint paper from Canada. Soon Rockefeller boasted of controlling 1,200 newspaper publishers by threatening their newsprint, which had to be carried on US ships.[16]

Rockefeller's media staff then saturated Latin America with planted news stories friendly to US and especially Rockefeller business interests in the region. Under the guise of fighting Nazi influence in Latin America, Nelson Rockefeller and his brothers were laying the basis of their vast private business empire for the postwar era.

Among the most far-reaching covert operations carried out by Nelson and his circle in Latin America towards the end of the War, was to secure for the United States the majority votes of participating nations in the founding of the United Nations, and with it, de facto US control of the International Monetary Fund and World Bank in 1944-45. It was indicative of how the new US international elite moved governments and others to suit their agenda. The UN was to be their vehicle, as they saw it, wrapped in the clothing of world democracy.

According to historian John Loftus, Rockefeller used behind-the-scenes pressure to get the backing of all Latin American nations in the founding San Francisco conference of the United Nations in 1945. This included the pro-Axis regime of Juan Peron in Argentina. Rockefeller and Washington pressured Peron to officially declare war on Germany and Italy, even though it was two weeks before the war's end. That allowed Argentina to vote with the "winning" side.

Rockefeller's political strategy was to use his block of Latin American nations to "buy" the majority vote at the UN. The Latin American bloc represented nineteen votes to Europe's nine. As a result, Washington and the powerful international banking business interests shaping its postwar agenda, ended up with decisive control of the IMF, the World Bank and a dominant role in the United Nations [17] The Rockefeller family, generous to a fault, even donated the land for the headquarters of the new United Nations in New York City. It was also good business, and a nice tax write-off to boot.

On the whole, Nelson Rockefeller was well situated in 1941, more than perhaps anyone else in US business circles, to launch his major Latin American agribusiness initiative.

The Rockefeller-Wallace Report

In 1941, some months before Pearl Harbor had brought the United States into the war, Rockefeller and US Vice President Henry A. Wallace, the former Agriculture Secretary of Franklin Roosevelt, sent a team to Mexico to discuss how to increase food production with the Mexican government. Wallace was a well-known agriculturist, who had served as Roosevelt's Secretary of Agriculture until 1940, and who had founded the seed company that became Pioneer Hi-Bred International Inc., which decades later would become a DuPont company and one of the Big Four GMO seed giants.

The Wallace-Rockefeller team's Mexico report emphasized the need to breed crops that had higher yields. At the time, corn was the major crop of Mexico, along with wheat and beans. In 1943, as a result of the project, the Rockefeller Foundation started the Mexican Agricultural Program (MAC), headed by the Rockefeller Foundation's George Harrar. The program included a young plant pathologist from the Rockefeller Foundation named Norman Borlaug. The Rockefeller family was preparing the first steps of what was to become a major transformation of world agricultural markets after the war.

That same year, as Nelson and Vice President Wallace were surveying Latin America for agricultural opportunities for the United States, Laurance and Nelson Rockefeller both had begun buying up, on the cheap, vast holdings of high-quality Latin American farmland. The family was diversifying their fortune from oil into agriculture.[18]

This was not simple family farming, however, but global "agribusiness," as it began to be called in the 1950's. Oil was the core of the new agribusiness economics. And oil was something the Rockefellers knew cold. The economic model of global monopoly concentration they had built up in oil over decades would be the model for transforming the nature of world agriculture into a global "agribusiness."

In March 1941, nine months before the bombing of Pearl Harbor brought the US into the war, Laurance took advantage of British financial duress in the Americas and bought 1.5 million acres of prime agricultural land on the Magdalena River in Colombia. Brother Nelson had also just bought a vast ranch in Venezuela, once owned by Simon Bolivar. As a Rockefeller aide at CIAA glibly said at the time, "There are good properties in the British portfolio. We might as well pick them up now."[19]

By the time Roosevelt named thirty-two-year-old Nelson Rockefeller to be Assistant Secretary of State for Latin America, Rockefeller was fully involved with food and agribusiness. In 1943, Edward O'Neal, President of the American Farm Bureau Federation, joined with Nelson and other top US businessmen at Chapultepec, Mexico, for a conference on Inter-American cooperation organized by the US State Department.

At Chapultepec, Rockefeller agreed with O'Neal that US agriculture needed new export markets. The markets of Latin America were coming into their view. Nelson said he was looking for new "frontiers." Rockefeller, in a true spirit of free market, demanded that the Americas be closed to all but US business interests, while demanding that the world, including governments of Latin America, open their doors to US products, including agriculture.[20]

Rockefeller also agreed with US Pentagon generals at Chapultepec, that selling US military surplus weapons to the governments of Latin America would be a good way to lock those countries into dependence on Washington for their military security after the war.[21] Dependence on US military security was to work in tandem with Latin American economic dependence on US companies and on US bank capital. No one was more in the forefront of this transition in the 1940's than the Rockefeller family. They also held major stock in the largest military defence industries.[22]

As the Cold War escalated in the late 1940's, Truman announced that the US would fight the expansion of Communism in Africa, Asia, and Latin America. He called for exporting US technical expertise and capital to the developing world, stressing that the American private sector, and not the US Government, should play the leading role in transferring US technology abroad.

The concept came from Nelson Rockefeller. US domination of global agricultural technology was fast becoming a weapon of the Cold War for Washington, and above all for the powerful Rockefeller interests.

By the beginning of the 1950's, US export of agricultural products was nearly equal in importance to arms and manufactures export. The US Department of Agriculture food surplus was viewed as a weapon of US foreign policy. As noted earlier, by 1954, P.L. 480 or "Food for Peace," had formalized the process in a major way.

The Rockefeller family and the Rockefeller Foundation had little problem getting their view of global food and population issues across to the US State Department. They and their allies from the New York Council on Foreign Relations dominated the senior ranks of the US foreign policy establishment.

The Rockefeller group wielded tremendous influence on the State Department. Every man who served as Secretary of State in the critical Cold War years ranging from 1952 to the end of Jimmy Carter's Presidency in 1979 had formerly been a leading figure from the Rockefeller Foundation.

Eisenhower's Secretary of State, John Foster Dulles, a Wall Street lawyer, was Chairman of the Rockefeller Foundation before he came to Washington in 1952. John Kennedy's and later Lyndon Johnson's Secretary of State, Dean Rusk, left his job as President of the Rockefeller Foundation to come to Washington in 1961. Nixon's National Security Adviser and Rusk's successor in 1974 as Secretary of State, Henry Kissinger, also came from the inner circle of the Rockefeller Foundation. Moreover, Jimmy Carter's Secretary of State, Cyrus Vance, came to Washington from his post as Chairman of the Rockefeller Foundation. But the enormous influence of this private, non-profit foundation on post-war American foreign policy was kept well in the background.

Dulles, Rusk, Vance and Kissinger all understood the Rockefeller views on the importance of private sector activity over the role of government, and they understood how the Rockefellers viewed agriculture—as a commodity just like oil, which could be traded, controlled, made scarce or plentiful depending on foreign policy goals of the few corporations controlling its trade.

Remarkably enough, the Dulles-Rusk-Vance-Kissinger-Rockefeller ties were rarely mentioned openly, even though they were essential to understanding key aspects of US foreign policy and food policy.

Early Agribusiness: Rockefeller Teams up with Cargill

In 1947, after the end of the War, Nelson Rockefeller founded another new company called International Basic Economy Corporation (IBEC). IBEC's aim was to show that private capital, organized as a profit-making enterprise, could upgrade the agriculture of developing countries. In reality, IBEC was about the introduction of mass-scale agribusiness in countries where US dollars could buy huge influence in the 1950's and 1960's.

Rockefeller's IBEC invited Cargill, a privately-held US agribusiness giant, to work with it in Brazil. IBEC had lots of plans: hybrid corn production, hog production, crop dusting with helicopters, contract plowing and grain storage. One IBEC company was Sementes Agroceres, which later played a key role in plant and animal genetics in Brazil.[23]

IBEC and Cargill began developing hybrid corn seed varieties. They turned Brazil into the world's third largest corn producer after the US and China. In Brazil, the corn was mixed with soymeal for animal feed. That was later to become instrumental in the proliferation of GMO soybeans on the world animal feed market in the late 1990's.

The agricultural economics of sugarcane also led to Brazil's prominent role in the production of soybeans. Sugarcane plants could typically produce for about five years after which they had to be dug out and new cane planted, a procedure known as "rationing." Brazilian farmers pioneered the planting of soybeans between the digging out of the old and planting of the new cane. Soybeans enriched or "fixed" nitrogen in the soil. Since sugarcane needs nitrogen, this reduced demand for fertilizer, which was the reason soybeans were introduced in Brazil.

Cargill and the other US grain trading companies later developed soybeans into a major export commodity, initially as animal feed. It became a major weapon in the US food control arsenal.

Lester Brown, whose own Worldwatch Institute was created with a 1974 grant from the Rockefeller Brothers Fund, stated the agenda of the Rockefeller Foundation's Green Revolution: "Fertilizer is in the package of new inputs which farmers need in order to realize the full potential of the new seed. Once it becomes profitable to use modern technology, the demand for all kinds of farm inputs increases rapidly. And so, only agro-business firms can supply these inputs efficiently."[24]

Brown further declared that the multinational corporation was "an amazingly efficient way of institutionalizing the transfer of technical knowledge in agriculture." And the agribusiness firms which were then in the best position to provide seeds and fertilizer were, of course, American agribusiness firms like DuPont, Pioneer Hi-Bred International, Cargill and Archer Daniels Midland. Thus, encouraged by the Rockefeller Green Revolution, beginning in the late 1950's, US agribusiness export was rapidly becoming a strategic core of US economic strategy alongside oil and military hardware.

In Brazil and Venezuela

As the Rockefeller Foundation's Green Revolution was making major inroads in Mexico, Nelson Rockefeller set up another organization to pursue similar work in Brazil and Venezuela. He wanted to continue projects he had started at the Office of the Coordinator of Inter-American Intelligence Affairs (CIAA) during World War II. Joining with several former CIAA colleagues, he created the American International Association for Economic and Social Development (AIA). The declared objective of the AIA was the transfer of technology and education.

With the AIA, Rockefeller wanted to rapidly modernize basic infrastructure. The AIA argued that if their efforts failed, the region faced the prospect that an exploding population would decrease the standard of living. As a major stockholder in Venezuela's Creole Petroleum, Rockefeller convinced Shell, Mobil, Gulf, and various other private donors to join him in underwriting the AIA's projects after 1946. Nelson and his brothers had sponsored a series of studies, a precursor to NSSM 200, pinpointing which nations in Latin America, Southeast Asia, the Middle East and Africa were likely to be "soft on communism." Brazil and Venezuela in Latin America were singled out in the study—Brazil because of its vast untapped wealth and Venezuela because of the Rockefeller family's involvement with its oil.[25]

Nelson A. Rockefeller was a master of deploying the rhetoric of Cold War necessity in the name of US "national security" while advancing family interests. It did not hurt his effort that his old friend and former head of the Rockefeller Foundation, John Foster Dulles, now Secretary of State, pursued a policy of nuclear "massive retaliation" and Cold War "brinksmanship," which made the population ever-aware of the alleged dangers and threat of the

Soviet military. That made it quite easy to justify almost anything in the name of "US national security interests."

What Nelson Rockefeller and other leading US bankers and businessmen were creating with agriculture in Latin America was the early phase of what was to be a revolution in world food production. In the process, they set out to take over the control of basic daily necessities of the majority of the world's population. Like most revolutions, it wasn't what it advertised itself to be.

The Rockefeller Foundation, not surprisingly, was at the forefront here too. They even gave the process a new term—agribusiness. Their model of agribusiness, driven by rules set out by the dominant player, US industry and finance, provided the perfect partner for the introduction, by the 1990's, of genetically engineered food crops or GMO plants. How this marriage of strategic interests came about and of what its longer-term goals consisted were to remain hidden under the rubric of free market efficiency, modernization, feeding a malnourished world and other public relations fabrications—thus cleverly obscuring the boldest coup over the destiny of entire nations ever attempted.

Notes

1. Peter Grose, *Continuing the Inquiry: The Council on Foreign Relations from 1921 to 1996*, New York, Council on Foreign Relations Press, 1996. pp. 23-26. This official account by the CFR of the War and Peace Studies project states: "More than two years before the Japanese attack on Pearl Harbor, the research staff of the Council on Foreign Relations had started to envision a venture that would dominate the life of the institution for the demanding years ahead. With the memory of the Inquiry in focus, they conceived a role for the Council in the formulation of national policy. On September 12, 1939, as Nazi Germany invaded Poland, (CFR's Hamilton Fish) Armstrong and Mallory entrained to Washington to meet with Assistant Secretary of State George S. Messersmith. At that time the Department of State could command few resources for study, research, policy planning, and initiative; on such matters, the career diplomats on the eve of World War II were scarcely better off than had been their predecessors when America entered World War I. The men from the Council proposed a discreet venture reminiscent of the Inquiry: a program of independent analysis and study that would guide American foreign policy in the coming years of war and the challenging new world that would emerge after. The project became known as the War and Peace Studies. "The matter is strictly confidential," wrote (Isaiah) Bowman, "because the whole plan would be 'ditched' if it became generally known that the State Department is working in collaboration with any outside group." The Rockefeller Foundation agreed to fund the project, reluctantly at first, but, once convinced of its relevance, with nearly $350,000. Over the coming five years, almost 100 men participated in the War and Peace Studies, divided into four functional topic groups: economic and financial, security and armaments, territorial, and political. These groups met more than 250 times, usually in New York, over dinner and late into the night. They produced 682 memoranda for the State Department, which marked them classified and circulated them among the appropriate government departments."

2. *Ibid.*, pp. 10, 15.

3. U.S. Supreme Court, *U S V. U S Steel Corporation*, U.S. 417, 1920, p. 251.

4. Henry Luce, "The American Century," *Life*, 17 February 1941.

5. Handbook, The New York Council on Foreign Relations, *Studies of American Interests in the War and the Peace*, New York, 1939-1942, cited in Neil Smith, *American Empire: Roosevelt's Geographer and the Prelude to Globalization*, University of California Press, Berkeley, 2003, pp. 325-328..

6. Neil Smith, *op. cit.*, p. 287.

7. Joseph S. Nye Jr, "Propaganda Isn't the Way: Soft Power", *The International Herald Tribune*, 10 January 2003. Nye defines what he coined as "soft power":

"Soft power is the ability to get what you want by attracting and persuading others to adopt your goals. It differs from hard power, the ability to use the carrots and sticks of economic and military might to make others follow your will. Both hard and soft powers are important…but attraction is much cheaper than coercion, and an asset that needs to be nourished."

8. Kramer, Paul, "Nelson Rockefeller and British Security Coordination", *Journal of Contemporary History*, Vol. 16, 1981, pp. 77-81.

9. Charles Higham, *Trading with the Enemy: An Exposé of the Nazi-American Money Plot, 1933-1947*, Delacorte, New York, 1983, pp.53-54.

10. *Ibid.*, p. 56.

11. *Ibid.*, pp. 67-69.

12. William Stevenson, *A Man Called Intrepid*, Ballantine Books, New York, 1976, pp. 308-311.

13. Gerard Colby and Charlotte Dennett, *Thy Will Be Done: The Conquest of the Amazon-Nelson Rockefeller and Evangelism in the Age of Oil*, HarperCollins, New York, 1995, pp. 115-116.

14. Thomas O'Brien, *Making the Americas: U.S. Business People and Latin Americans from the Age of Revolutions to the Era of Globalization*, History Compass ?, LA 067, 2004, pp. 14 15

15. Los Angeles Times, *Mexico 75 Years Later, Today's Zapatistas Still Fight the Rockefeller Legacy*, 14 May 1995.

16. William Stevenson, *op. cit.*, p. 309.

17. John Loftus and Mark Aarons, *The Secret War Against the Jews: How Western Espionage Betrayed the Jewish People*, St. Martin's, New York, 1994, pp. 165-171.

18. Margaret Carroll Boardman, *Sowing the Seeds of the Green Revolution: The Pivotal Role Mexico and International Non-Profit Organizations Play in Making Biotechnology an Important Foreign Policy Issue for the 21st Century*, http://www.isop.ucla.edu/profmex/ volume4/3summer99/Green_Finalm.htm.

19. Gerard Colby and Charlotte Dennett, *op. cit.*, pp. 116, 168.

20. *Ibid.*, p. 166.

21. *Ibid.*, p. 169.

22. Committee on Rules and Administration, U.S. Senate, 93rd Congress, 2nd Session, Hearings, *The Nomination of Nelson A. Rockefeller of New York to be Vice President of the United States*, Washington D.C., Government Printing Office, 1974, cited in Gerard Colby and Charlotte Dennett, *op. cit.*, p. 373. In addition to the known holdings in the Standard Oil companies, Rockefeller's investments included such prime defense contractors as McDonnell Aircraft, Chrysler Corp.,

Boeing, Monsanto, Dow Chemical, Hercules, Bendix, Motorola and numerous other defense contractors.

23. John Freivalds, *Brazil Agriculture: Winning the Great Farms Race,* 3 March 2005, http://www.brazilmax.com/news.cfm/tborigem/fe_business/id/5.

24. Lester Brown, Seeds of Change, Praeger, New York, 1969, *Chapter 1: New Seeds and Mechanization.*

25. Gerard Colby and Charlotte Dennett, *op. cit.,* pp. 212-214.

PART III
Creating Agribusiness

CHAPTER 7
Rockefeller and Harvard Invent USA "Agribusiness"

A Green Revolution Opens the Door

The Rockefellers' Green Revolution began in Mexico and was spread across Latin America during the 1950's and 1960's. Shortly thereafter, backed by John D. Rockefeller's networks across Asia, it was introduced in India and elsewhere in Asia. The "revolution" was a veiled effort to gain control over food production in key target countries of the developing world, promoted in the name of free enterprise market efficiency against alleged "communist inefficiency."

In the aftermath of World War II, with Germany's I.G. Farben a bombed-out heap of rubble, American chemical companies emerged as the world's largest. The most prominent companies—DuPont, Dow Chemical, Monsanto, Hercules Powder and others—faced a glut of nitrogen production capacity which they had built up, at US taxpayer expense, to produce bombs and shells for the war effort.

An essential chemical for making bombs and explosives, nitrogen was a prime component of TNT and other high explosives. Nitrogen could also form the basis for nitrate fertilizers. The chemical industry developed the idea of creating large new markets for their nitrogen in the form of fertilizers, ammonia nitrate, anhydrous ammonia, for both domestic US agriculture and for export.

The nitrogen fertilizer industry was part of the powerful lobby of the Rockefeller Standard Oil circles which, by the end of the War, included DuPont, Dow Chemicals and Hercules Powder among others.

The global marketing of the new agri-chemicals after the war also solved the problem of finding significant new markets for the American petrochemical industry as well as the grain cartel, a group of four to five companies then including Cargill, Continental Grain, Bunge and ADM. The largest grain traders were American and their growth was a product of the development of special hybrid seeds through the spread of the Green Revolution in the 1960's and 1970's. Agriculture was in the process of going global and the Rockefeller Foundation was shaping that process of agribusiness globalization.

With a monopoly on the agricultural chemicals and on the hybrid seeds, American agribusiness giants were intent on dominating the global market in agricultural trade. After all, as Kissinger noted in the 1970's, "If you control the food you control the people." Governments from the developing sector to the European Economic Community, the Soviet Union and China, soon depended on the powerful grain cartel companies to provide the needed grains and food products to maintain their political stability in times of bad harvest.

Truly, there was genuine US Government concern to contain communist and nationalist movements in the developing world during the 1960's by offering food aid in the form of privately sponsored agricultural inputs. However, the combination of US Government aid and the techniques being developed in the name of a Green Revolution would present a golden opportunity for the influential policy-making circles around Rockefeller and their emerging agribusiness groups to turn that concern to their advantage.

Nelson Rockefeller worked hand-in-glove on agriculture with his brother, John D. III, who had set up his own Agriculture Development Council in 1953, one year after he had founded the

Population Council. The focus of the Agriculture Development Council was Asia, while Nelson concentrated on his familiar turf in Latin America. They shared the common goal of long-term cartelization of world agriculture and food supplies under their corporate hegemony.

When the Rockefeller Foundation's Norman Borlaug came into Mexico in the 1950's, he worked on hybrid forms of rust-resistant wheat and hybrid corn types, not yet the genetically engineered projects to come several decades later. Behind the façade of agricultural and biological science, however, the Rockefeller group was pursuing a calculated strategy through its Green Revolution during the 1950's and 1960's.

The heart of its strategy was to introduce "modern" agriculture methods to increase crop yields and, so went the argument, thereby to reduce hunger and lessen the threat of potential communist subversion of hungry, unruly nations. It was the same seducing argument used years later to sell its Gene Revolution.

The Green Revolution was the beginning of global control over food production, a process made complete with the Gene Revolution several decades later. The same companies, not surprisingly, were involved in both, as were the Rockefeller and other powerful US foundations.

In 1966, the Rockefeller Foundation was joined by the considerable financial resources of the Ford Foundation, another US private tax-exempt foundation which enjoyed intimate ties to the US Government, intelligence and foreign policy establishment. Together with the Ford resources, the Rockefeller Foundation's Green Revolution went into high gear.

That year of 1966, the Government of Mexico along with the Rockefeller Foundation set up the International Maize and Wheat Improvement Center (CIMMYT). The center focused its work on a wheat program, which originated from breeding studies begun in Mexico in the 1940s by the Rockefeller Foundation.[1]

Their efforts in food and agriculture received a boost that same year when US President Lyndon Johnson announced a drastic shift in US food aid to developing countries under P.L. 480, namely that no food aid would be sent unless a recipient country had agreed to preconditions which included agreeing to the Rockefeller agenda for agriculture development, stepping up their population control programs and opening their doors to interested American investors.[2]

In 1970, the Rockefeller's Norman Borlaug won the Nobel Prize. Interestingly enough, it was not for biology but for peace, the same prize Henry Kissinger was to receive several years later. Both men were also protégés of the influential Rockefeller circles.

In reality, the Green Revolution introduced US agribusiness into key developing countries under the cover of promoting crop science and modern techniques. The new wheat hybrids in Mexico required modern chemical fertilizers, mechanized tractors and other farm equipment, and above all, they required irrigation, which meant pumps driven by oil or gas energy.

The Green Revolution methods were suitable only in the richest crop areas, and it was deliberately aimed at the richest farmers, reinforcing old semi-feudal Latifundist divisions between wealthy landowners and poor peasant farmers. In Mexico, the new wheat hybrids were all planted in the rich, newly-irrigated farm areas of the Northeast. All inputs, from fertilizers to tractors and irrigation, required petroleum and other inputs from advanced industrial suppliers in the United States. Oil and agriculture joined forces under the Rockefeller aegis.

In India, the Green Revolution was limited to 20 percent of land in the irrigated North and Northwest. It ignored the huge disparity of wealth between large feudal landowners in such areas and the majority of poor, landless peasants. Instead, it created pockets of modern agribusiness tied to large export giants such as Cargill. The regions where the vast majority of poorer peasants worked remained poor. The introduction of the Green Revolution did nothing to change the gap between rich feudal landowners and poor peasants, but overall statistics showed significant rises in Indian wheat production.

Training Cadre for the Bio-Revolution

In 1960, the Rockefeller Foundation, John D. Rockefeller III's Agriculture Development Council and the Ford Foundation joined forces to create the International Rice Research Institute (IRRI) in Los Baños, the Philippines. By 1971, the Rockefeller Foundation's IRRI, along with their Mexico-based International Maize and Wheat Improvement Center and two other Rockefeller and Ford Foundation-created international research centers, the IITA for tropical agriculture, Nigeria, and IRRI for rice, Philippines, combined to form a global Consultative Group on International Agriculture Research (CGIAR).[3]

CGIAR was shaped at a series of private conferences held at the Rockefeller Foundation's conference center in Bellagio, Italy. Key participants at the Bellagio talks were the Rockefeller Foundation's George Harrar, Ford Foundation's Forrest Hill, Robert McNamara of the World Bank and Maurice Strong, the Rockefeller family's international environmental organizer, who, as a Rockefeller Foundation Trustee, organized the UN Earth Summit in Stockholm in 1972.

To ensure maximum impact, CGIAR drew in the United Nations' Food and Agriculture Organization (FAO), the UN Development Program (UNDP) and the World Bank. Thus, through a carefully-planned leverage of its initial funds, Rockefeller by the beginning of the 1970's was in a position to shape global agriculture policy.[4]

Financed by generous Rockefeller and Ford Foundation study grants, CGIAR saw to it that leading Third World agriculture scientists and agronomists were brought to the US to "master" the concepts of modern agribusiness production, in order to carry it back to their homeland. In the process, they created an invaluable network of influence for US agribusiness promotion in those countries, all in the name of science and efficient free market agriculture.

This Rockefeller Foundation network of institutes and research centers had gradually laid the basis for control over agricultural research and policy for much of the developing world by the time Kissinger was commissioned to draft NSSM 200.

John D. Rockefeller III's Agricultural Development Council also deployed US university professors to select Asian universities to train a new generation of scientists. The best scientists would then be selected to be sent to the United States to get their doctorate in agriculture sciences, and coming out of the American universities, would follow the precepts close to the Rockefeller outlook on agriculture. This carefully-constructed network was later to prove crucial in the Rockefeller Foundation's subsequent strategy to spread the use of genetically-engineered crops around the world.

In a widely read handbook, Arthur Mosher, Executive Director of the Rockefeller Agriculture Development Council, insisted on teaching peasants to "want more for themselves." They were to be urged to abandon "collective habits" and get on with the "business of farming." Rockefeller's Mosher called for extending educational programs for women and building youth clubs, to create more demand for store-bought goods. He argued that, the "affection of husbands and fathers for their families" would make them responsive to these desires and drive them to work harder. Of course they would have to take out loans to invest in all this new technology, tying them even more to the new market economy.[5]

Through the Green Revolution, the Rockefeller and Ford Foundations worked hand-in-hand with the foreign policy goals of the United States Agency for International Development (USAID) and of the CIA.

One major effect of the Green Revolution was to depopulate the countryside of peasants who were forced to flee into shantytown slums around the cities in desperate search for work. That was no accident; it was part of the plan to create cheap labor pools for forthcoming US multinational manufactures.

When the self-promotion around the Green Revolution died down, the true results were quite different from what had been promised. Problems had arisen from indiscriminate use of the new chemical pesticides, often with serious health consequences. The mono-culture cultivation of new hybrid seed varieties decreased soil fertility and yields over time. The first results were impressive:

double or even triple yields for some crops such as wheat and later corn in Mexico. That soon faded.[6]

The Green Revolution was typically accompanied by large irrigation projects which often included World Bank loans to construct huge new dams, and flood previously settled areas and fertile farmland in the process. Also, super-wheat produced greater yields by saturating the soil with huge amounts of fertilizer per acre, the fertilizer being the product of nitrates and petroleum, commodities controlled by the Rockefeller-dominated Seven Sisters major oil companies.

Huge quantities of herbicides and pesticides were also used, creating additional markets for the oil and chemical giants. As one analyst put it, in effect, the Green Revolution was merely a chemical revolution. At no point could developing nations pay for the huge amounts of chemical fertilizers and pesticides. They would get the credit courtesy of the World Bank and special loans by Chase Bank and other large New York banks, backed by US Government guarantees.

Applied in a large number of developing countries, those loans went mostly to the large landowners. For the smaller peasants the situation worked differently. Small peasant farmers could not afford the chemical and other modern inputs and had to borrow money. Initially various government programs tried to provide some loans to farmers so that they could purchase seeds and fertilizers.

Farmers who could not participate in this kind of program had to borrow from the private sector. Because of the exorbitant interest rates for informal loans, many small farmers did not even get the benefits of the initial higher yields. After harvest, they had to sell most if not all of their produce to pay off loans and interest. They became dependent on money-lenders and traders and often lost their land. Even with soft loans from government agencies, growing subsistence crops gave way to the production of cash crops.[7]

The Green Revolution also introduced new machines for land preparation. Most notable was the so-called power tiller or turtle tiller. This machine, which puddled the rice paddy soil, also destroyed much of the natural soil structure. But, it was very efficient in doing that.

Another crucial aspect driving the interest of US agribusiness companies was the fact that the Green Revolution was based on proliferation of new hybrid seeds in developing markets. One vital aspect of hybrid seeds was their lack of reproductive capacity. Hybrids had a built in protection against multiplication. Unlike normal open pollinated species whose seed gave yields similar to its parents, the yield of the seed borne by hybrid plants was significantly lower than that of the first generation.

That declining yield characteristic of hybrids meant farmers must normally buy seed every year in order to obtain high yields. Moreover, the lower yield of the second generation eliminated the trade in seed that was often done by seed producers without the breeder's authorization. It prevented the redistribution of the commercial crop seed by middlemen. If the large multinational seed companies were able to control the parental seed lines in house, no competitor or farmer would be able to produce the hybrid. The global concentration of hybrid seed patents into a handful of giant seed companies, led by Pioneer HiBred and Monsanto's Dekalb laid the ground for the later GMO seed revolution.[8]

In effect, the introduction of modern American agricultural technology, chemical fertilizers and commercial hybrid seeds all made local farmers in developing countries, particularly the larger more established ones, dependent on foreign inputs. It was a first step in what was to be a decades-long, carefully planned process. Agribusiness was making major inroads into markets which were previously of limited access to US exporters. The trend was later dubbed "market-oriented agriculture." In reality it was agribusiness-controlled agriculture.

The Green Revolution and its hybrid seeds promised a major new controlled market for US agribusiness. Henry Wallace, Franklin Roosevelt's Secretary of Agriculture, had built the first major hybrid seed company, Pioneer Hi-Bred, largely by encouraging selective USDA government research on the positive yield gains of hybrids and downplaying their negative features. It enabled the growth of huge commercial seed companies. This laid the basis for the later development of genetic patented seeds by a handful of Western agribusiness giants.

The chemical industry also claimed that the increased crop yields were only possible with the help of their products. The US Government, through US AID and other government aid programs, backed this view, and convinced the host developing sector governments to support them. This led to a situation where farmers disregarded other more traditional means of yield improvement, which were labeled primitive and inefficient by the Rockefeller and Ford country advisers.[9]

Use of High Yield Varieties (HYV) of hybrid wheat, corn or rice, and major chemical inputs soon became the dominant practice. Local government officials no longer considered the option of possible yield improvement based on traditional practices. Often, the international chemical industry intervened to suppress or hinder research programs that would challenge their high input approach. This was a worldwide trend.[10]

In 1959, a team led by the US Department of Agriculture published the Ford Foundation's *Report on India's Food Crisis and Steps to Meet It*. In place of fundamental changes such as redistribution of land and other rural assets from the large quasi-feudal landowners as the foundation for a more effective Indian agricultural development, the Ford report stressed technological change including improved seeds, chemical fertilizers, and pesticides in small, already irrigated pockets of the country. It was the "Green Revolution" strategy.

Ford even funded India's Intensive Agricultural Development Program (IADP) as a test case of the strategy, providing rich farmers in irrigated areas with subsidized inputs, generous credit and price incentives. The World Bank backed the strategy with generous loans.

Soon, the Rockefeller-Ford Green Revolution was adopted by the Indian government, with far-reaching effects. Agricultural production of rice and wheat in the selected pockets grew immediately with the new hybrids and chemical inputs. Talk of land reform, tenancy reform, abolition of usury, was dropped from the official Indian Government agenda, never to return.[11]

The initial spectacular growth rates eventually slowed, though this aspect was not widely publicized, leaving the one-sided impression of success. On average, overall agricultural production in India grew more slowly after the Green Revolution than before, and in much of the country, per capita agricultural output stagnated or fell.[12] But the Green Revolution had one success: it created a large new market for US and foreign agribusiness multinational firms to sell their chemicals, petroleum, machinery and other inputs to developing countries. It was the beginning of what was called agribusiness.

Rockefeller Finances the Creation of Agribusiness

While the Rockefeller brothers were expanding their global business reach from oil to agriculture in the developing world through their Green Revolution scheme, they were financing a little-noticed project at Harvard University, which would form the infrastructure to globalize world food production under the central control of a handful of private corporations. Its creators gave it the name "agribusiness," in order to differentiate it from traditional farmer-based agriculture—that is, the cultivation of crops for human sustenance and nutrition.

Agribusiness and the Green Revolution went hand-in-hand. They were part of a grand strategy which included Rockefeller Foundation financing of research for the development of genetic alteration of plants a few years later.

John H. Davis had been Assistant Agriculture Secretary under President Dwight Eisenhower in the early 1950's. He left Washington in 1955 and went to the Harvard Graduate School of Business, an unusual place for an agriculture expert in those days. He had a clear strategy. In 1956, Davis wrote an article in the *Harvard Business Review* in which he declared that "the only way to solve the so-called farm problem once and for all, and avoid cumbersome government programs, is to progress from agriculture to agribusiness." He knew precisely what he had in mind, though few others had a clue back then.[13]

Davis, together with another Harvard Business School professor, Ray Goldberg, formed a Harvard team with the Russian-born economist, Wassily Leontief, who was then mapping the entire US economy, in a project funded by the Rockefeller Foundation. During the war, the US Government had hired Leontief to develop a method of inter-sectoral analysis of the total economy which he referred to as input-output analysis. Leontief worked for the US Labor Department as well as for the Office of Strategic Services (OSS), the predecessor of the CIA.[14]

In 1948, Leontief got a major four-year $100,000 grant from the Rockefeller Foundation to set up the Harvard "Economic Research Project on the Structure of the American Economy." A year later, the US Air Force joined the Harvard project, a curious engagement for one of the prime US military branches. The transistor and electronic computers had just been developed along with methods of linear programming that would allow vast amounts of statistical data on the economy to be processed. Soon the Ford Foundation joined in the Harvard funding.

The Harvard project and its agribusiness component were part of a major attempt to plan a revolution in US food production. It was to take four decades before it dominated the food industry. Goldberg later referred to the agribusiness revolution and the development of gene-modified agribusiness as "changing our global economy and society more dramatically than any other single event in the history of mankind."

Monopoly and Vertical Integration Return with a Vengeance

As Ray Goldberg boasted years later, the core idea driving the agribusiness project was the re-introduction of "vertical integration" into US food production. By the 1970's, few Americans realized that bitter battles had been fought to get Congress to outlaw vertical integration by giant conglomerates or trusts such as Standard Oil, in order to prevent them from monopolizing whole sectors of vital industries.

It wasn't until the David Rockefeller-backed Presidency of Jimmy Carter in the late 1970's that the US multinational business establishment was able to begin the rollback of decades of carefully constructed US Government regulations of health, food safety and consumer protection laws, and open the doors to a new wave of vertical integration. The vertical integration process was sold to unaware citizens under the banner of "economic efficiency" and "economy of scale."

A return to vertical integration and the accompanying agribusiness were introduced amid a public campaign in prominent media claiming that government had encroached far too much into the daily lives of its citizens and had to be cut back to give ordinary Americans "freedom." The war cry of the campaigners was "deregulation." What they carefully left out of their propaganda was that deregulation by government merely opened the door to de facto private regulation by the largest and most powerful corporate groups in a given industry.

The person who first called openly for deregulation of government controls and privatization, well before Jimmy Carter, Ronald Reagan or Margaret Thatcher, was John D. Rockefeller III. In 1973, he published *The Second American Revolution*. In the book and in numerous public addresses, Rockefeller called for a "deliberate, consistent, long-term policy to decentralize and privatize many government functions ... to diffuse power throughout the society."[15]

Well before that, however, Davis and Goldberg had begun to industrialize specific sectors of American agriculture into agribusiness through vertical integration, ignoring anti-trust laws, and using Leontief's input-output approach to identify the entire production and distribution chain.

The first result of the collaboration between Davis, Goldberg and Leontief was a project to industrialize the Florida citrus industry. The control of small citrus farmers soon gave way to large national orange juice processors such as Sunkist, who dominated prices paid to the farmer through control of distribution and processing.[16]

Their next target was to develop a strategy for the industrialization of the US wheat-to-consumer chain as well as the soybean market for animal feed. As the Government, step-by-step, removed regulatory controls on agriculture or on monopoly, the vertical integration of the food industry accelerated.

Significantly, the first American industry to be completely vertically integrated had been oil, under the Rockefeller Standard Oil Trust in 1882. Despite repeated attempts by numerous states to outlaw Rockefeller's monopolistic control of oil and freight prices, even a Supreme Court decision in 1911 failed to break up the cartel in oil, which went on to dominate the global oil trade for the following century. The Standard Oil model, not surprisingly, was the model for the Harvard Rockefeller Foundation project to create agribusiness from agriculture.

In the 1920's, a series of laws had been passed by the US Congress to control food monopolies, especially in the meat sector, following the revelation of shocking practices in the US meatpacking and processing industry, by writers such as Upton Sinclair whose book *The Jungle* described the fetid, unsanitary and often inhuman conditions of the meatpacking industry.

Five major companies—Armour, Swift, Morris, Wilson and Cudahy—were then in a position, as the US Government's newly-founded Federal Trade Commission (FTC) accused them, of trying "to monopolize all the nation's food supply" by the 1920's. The five had systematically and illegally acquired a near monopoly in meatpacking.[17]

The Big Five then controlled who had access to public stockyards for the cattle. They interfered with the livestock marketing process through monopoly control, controlled wholesale distribution channels, and restricted what retailers could buy. With the invention of the refrigerated railcar and assembly-line continuous meat processing plants, the meat companies vertically integrated. They integrated forward into marketing the beef, and backward into monopolizing supply of raw material—beef cattle and hogs.

An FTC investigation in the early 1920's found that the five companies had dominated the purchase of livestock by controlling major stockyards, terminal railroads, livestock credit, market news media, and sites for potential rival packing plants. Furthermore, they had used their domination to force out new competitors and had cartelized the remaining market among themselves illegally. They controlled the retail level by owning refrigerator transport cars, cold storage warehouses and severely reduced competitor market access. Not content with all that, according to the Government investigation, the Big Five meat packers also controlled the market for substitute foods by buying or controlling them.[18]

By the 1970's, the US food supply was once more going into the hands of a tiny, monopoly of agribusiness producers. This time, aided by the Rockefeller and Ford Foundation funding of the Harvard Economic Research Project on the Structure of the American Economy under Leontief, Goldberg and Davis were spearheading a new corporate rush into vertical integration and monopoly control of not only American but global food supply. The scale was without precedent.

Goldberg and Davis and their colleagues at Harvard were at the forefront of educating a new generation of corporate managers who would be infected with the prospect of staggering profits in the effort to totally restructure the way Americans grew food to feed themselves and the world.

As US Government regulatory barriers fell under the drumbeat of deregulation, especially during the Presidency of Ronald Reagan, agribusiness rushed in to fill the regulation vacuum with its own private industry rules and standards. The standards were not set by all players, but typically rather by the top four or five monopoly players.

The process led to a concentration and transformation of American agriculture. Independent family farmers were driven off the land to make way for "more efficient" giant corporate industrial farm businesses, known as Factory Farms or corporate agriculture. Those who stayed on the land were mostly forced to work for the big agribusiness firms as "contract farmers."

"Where Have all the Farmers Gone?"
As Government regulations, food safety standards and monopoly laws were systematically loosened, especially during the 1980's

Reagan-Bush era, agribusiness began to transform the face of traditional American farming in ways so drastic as to be incomprehensible to ordinary consumers. Most people simply went to their local supermarket, took a nicely packed cut of beef or pork from the meat counter and thought they were still buying the product of the family farm.

What began to take place instead was the wholesale merger and consolidation, one-by-one, of American food production, out of the hands of family farmers and into giant corporate global concentrations. The farmer gradually became a contract employee responsible only for feeding and maintaining concentrations of thousands of animals in giant pens. He no longer owned the animals or the farm. He was effectively becoming like a feudal serf, indentured through huge debts, not to a Lord of the manor, but to a global multinational corporation such as Cargill, Archer Daniels Midland, Smithfield Foods or ConAgra.

For the new corporate agribusiness giants, the transformation was quite profitable. Family farmers' income for the vast majority of farm families plunged as they lost control of their market entirely to the agribusiness giants by the end of the 1990's. Their returns on equity had fallen from an average of 10% in the mid-1970's to only 2% a year, according to a study by the Senate Agriculture Committee. At the same time, the average annual return on stockholder equity for the industrialized food processing sector rose to 23% by 1999 from 13% in 1993.[19]

Hundreds of thousands of independent family farmers were forced out of business with the spread of agribusiness and its large operations. They simply couldn't compete. Traditional farming was by its nature labor intensive, while factory farming was capital intensive. Farmers who did manage to raise the money for animal confinement systems quickly discovered that the small savings in labor costs were not enough to cover the increasing costs of facilities, energy, caging, and drugs.

The increase in factory farms led to a decrease in the price independent farmers got for their animals, forcing thousands out of business. The number of US farmers dropped by 300,000 between 1979 and 1998.[20]

The number of hog farms in the US decreased from 600,000 to 157,000, while the number of hogs sold increased. Consolidation resulted in just 3 percent of US hog farms producing more than 50 percent of the hogs. A report to the US Secretary of Agriculture in the late 1990's described the enormous social costs of the destruction of the American family farm by agribusiness, as the economic basis of entire rural communities collapsed and rural towns became ghost towns. The USDA report was buried.[21]

Another minority report led by Senator Tom Harkin, released just before the November 2004 US Presidential elections, and also buried, revealed that by then the degree of concentration and near-monopoly in the food and agriculture economy of the United States was impressive to say the least. The report found that the four largest beef packers controlled 84% of steer and heifer slaughter and 64% of hog slaughter. Four companies controlled 89% of the breakfast cereal market.[22]

When Cargill acquired the grain handling operations of Continental Grain in 1998, that one company, Cargill controlled 40% of national grain elevator capacity. The US Justice Department approved the merger. Four large agro-chemical/seed companies—Monsanto, Novartis, Dow Chemical, and DuPont—control more than 75 percent of the nation's seed corn sales and 60 percent of soybean seed sales, at the same time that these companies control large shares of the agricultural chemical market.[23]

As traditional farmers abandoned their family land in droves during the 1980's and 1990's, agribusiness moved in to fill the void. The extent of the dramatic shift was largely hidden by clever government statistical accounting methods to make it appear that family farmers were simply getting larger, not that American farming had become giant corporate agribusiness.[24]

Municipalities, often desperate to attract jobs in regions of rural depression, offered the new agribusiness giants attractive concessions, tax benefits and others, to locate their industrial farms in the region, hoping to create new jobs and economic growth. The main growth created by the huge animal concentrations was fecal matter—animal waste in unimagined volumes.

What was termed a revolution in animal factory production began in the early 1980's. It was unpublicized for obvious reasons. Techniques of mass production and factory efficiency were introduced by the large corporations much as had been done in the auto industry assembly line production. Hogs, cattle and chickens were no longer produced on open fields or small farms where animals received individual attention from the farmer in event of illness or disease. The new production involved what was called "confinement feeding" or what came to be called CAFOs—Concentrated Animal Feeding Operations. Their goal was maximum corporate profit at minimum cost—Shareholder Value was the Wall Street term. Gone was a system in which direct attention and care to the individual pig or cow or pasture land or crop soil mattered. Profit was the bottom line of the corporate agribusiness giant driving the transformation.

The CAFOs brought impressive concentrations of animal flesh into the smallest possible confinement space. From birth to slaughter, a factory pig, often weighing 500 to 600 pounds, would never leave a typical gestation cage of concrete and bars, a cell only as large as the animal. The animal would never be able to lie down, and as a result developed severe foot problems. The unnatural confinement created madness in the sows, including "bar biting" and senseless chewing. Never in their entire life did they see daylight.

The US Department of Agriculture estimated that 10% of all animals confined in CAFOs died annually due to stress, disease and injury, and up to 28% for some types of chickens. The factory managers had no incentive to spend time or invest in individual animals, arguing that it was more "cost effective" to take some "loss on inventory" rather than invest in proper veterinary care. Factory farming, as a result of generous campaign contributions to Congressmen, enjoyed an exempt status from normal laws against cruelty to animals.[25]

Cattle were packed into similar cages by the thousands. The London *Economist* magazine, in a May 2000 report, described the transformation of Iowa into the largest pig production center in America under factory farming. "Take take a trip to hog heaven," they wrote. "This ten-mile stretch of countryside north of Ames, Iowa, produces almost a tenth of America's pork. But there is not an animal in sight. In massive metal sheds, up to 4,000 sows at a time are reared for slaughter, their diets carefully monitored, their waste regularly siphoned away, their keepers showered and be-gowned, like surgeons, to avoid infecting the herd."[26]

OMB Watch, an organization monitoring the role of US Government regulators in the area, reported the effects of the drastic reduction in Government rules on pollution and animal waste contamination from giant factory farm installations beginning during the Carter Presidency in the seventies.

Under the George W. Bush Administration, the Environmental Protection Agency, at the request of agribusiness, repealed a rule that held corporate livestock owners liable for damage caused by animal waste pollution. They noted that the factory farm owners often evaded responsibility by hiring contractors to raise their animals. The EPA also dropped a requirement that would have forced facilities to monitor groundwater for potential contamination by animal waste, which often seeped into the earth, leaving communities vulnerable to potentially dangerous drinking water supplies. The EPA had refused to change the allowed levels of which livestock operations met their definition of CAFO with attendant pollution limits despite repeated lawsuits.[27]

Because of the huge scale of the CAFOs or Factory Farms, animal waste and pollution of ground water was no minor affair. The huge animal farms housed tens of thousands of cattle, pigs or chickens in small concentrations, hence the name, CAFO. It was estimated that the factory farms produced more than 130 times the waste that humans did, or some *2.7 trillion pounds of animal waste* a year.[28] That waste would then be channelled into enormous "lagoons" that often leaked, ruptured or overflowed—killing fish and other marine life, spreading disease and contaminating community drinking water supplies. The CAFO farms also routinely over-applied liquid waste to land areas, known as "sprayfields," causing it to run into waterways.

"Water contaminated by animal manure contributes to human diseases such as acute gastroenteritis, fever, kidney failure, and even death," according to a 2005 study by NRDC.[29]

Among the findings documented by the NRDC study were some alarming consequences to the cartelization of US agribusiness. They documented that in 1996 the US Government's Centers for Disease Control established a link between spontaneous abortions and high nitrate levels in Indiana drinking water wells located close to animal feedlots. As well, the high levels of nitrates in drinking water also increase the risk of methemoglobinemia, or "blue-baby syndrome," which can kill infants. Further, animal waste contains disease-causing pathogens, such as Salmonella, E. coli, Cryptosporidium, and fecal coliform, which can be 10 to 100 times more concentrated than in human waste. More than 40 diseases can be transferred to humans through manure.[30]

Typically, the corporations running the CAFOs would hire illegal immigrants at dirt low wages to deal with the huge waste concentrations, channelling them into vast "lagoons" which often ruptured or overflowed, killing fish and contaminating drinking water supplies.[31]

By the end of the 1990's, factory farming had made agriculture into the United States' largest general source of water pollution. One study showed that a growing hog produced two to four times as much waste as a human and a milk cow the waste of 24 people. Spread over large fields in a traditional family farm, such waste had never been a serious ecological problem. Concentrated into industrial centers of maximum animal density per square foot, it created staggering new environmental and health hazards. Because of the financial muscle of the giant corporate agribusiness farms, the Government catered to their needs to maximize profits, ignoring their legislative mandate to guard public health.

To deal with the large manure problem, the CAFOs typically would build earth pits to hold tens of millions of gallons of festering manure with an estimated "pollution strength" 130 times greater than human sewage. Putrid manure and urine waste contaminated countless streams and ground water sources across the United States.[32]

In California's Central Valley, giant mega-dairy CAFOs, with a total of 900,000 dairy cows, leaked fecal matter into the ground water, pushing nitrate levels of drinking water up 400%. The waste produced by the animals was equivalent to that of 21 million people.[33]

Not only waste, but consumption of drugs, especially antibiotics to keep diseases under control in the concentrated breeding spaces, became staggering. By the end of the 1990's the largest users of antibiotics and similar drugs from the large pharmaceutical firms were not humans, but animals, who consumed 70% of all pharmaceutical antibiotics.[34] The big pharmaceutical industry was becoming an integral part of the agribusiness chain.

In 1954, just as Harvard's Goldberg and Davis were developing their ideas on agribusiness, American farmers used about 500,000 pounds of antibiotics a year raising food animals. By the year 2005, it had increased to 40 million pounds, an eighty-fold rise. And some 80% of the antibiotics were poured directly into the animal feed to make the animals grow faster. Penicillin and tetracycline were the most commonly used antibiotics on the factory farms.

One result was the evolution of new strains of virulent bacteria appearing in humans and resistant to antibiotics. The Center for Disease Control and the USDA reported that the spread of food-related disease in humans resulting from eating meat pumped with antibiotics and other substances was "epidemic." Most of the food-related diseases were caused by contamination of the food, milk or water from animal fecal matter.[35]

The ability for corporations to merge and vertically integrate created a corporate concentration never before seen in agriculture. By the end of the 1990's, four large corporations—Tyson, Cargill, Swift and National Beef Packing—controlled 84% of all beef packing in the United States. Four corporations—Smithfield Foods, Tyson, Swift and Hormel—controlled 64% of all pig packing. Cargill, ADM and Bunge controlled 71% of all soybean crushing, and Cargill, ADM and ConAgra controlled 63% of all flour milling. Two GMO giants, Monsanto and Pioneer-HiBred of Dupont controlled 60% of the US corn and soybean seed market, which consisted entirely of patented Genetically Modified seeds. The ten largest food retailing corporations, led by Wal-Mart, controlled a total global market of $649 billion by 2002.[36]

By the beginning of the new millennium, corporate agribusiness had vertically integrated into a concentration of market power never before experienced even in the trust heyday of the early 1920's. Agribusiness as a sector had become the second most profitable industry in America next to pharmaceuticals, with annual domestic sales of well over $400 billion.[37] And the next phase was clearly mergers between the pharmaceutical giants and the agribusiness giants.

It was not surprising that the Pentagon's National Defense University, on the eve of the 2003 Iraq war, issued a paper declaring: "Agribusiness is to the United States what oil is to the Middle East."[38] Agribusiness had become a strategic weapon in the arsenal of the world's only Superpower.

The giant factory farms also destroyed the viability of traditional farming, killing an estimated three traditional farm jobs for every new, often low-paid, job it created. Shareholder Value had come to American agriculture with a vengeance.

The United States Department of Agriculture had been established in 1862 by President Abraham Lincoln who called it "the peoples" department. Its original mandate had been to serve farmers and their families, about half the population of the country at the time. By the end of the 20th Century, the number of family farmers had been decimated. The traditional farmer had become a near extinct species under the driving pressures of agribusiness and its power to control entire sectors through vertical integration.

The US Department of Agriculture or USDA had been transformed into a lobby for agribusiness. Between 1995 and 2003 American taxpayers paid over $100 billion for USDA crop subsidies. The subsidies went not to struggling family farmers, however. They went overwhelmingly to the giant new agribusiness operators, corporate farms, including millions to David Rockefeller, the ardent advocate of less government subsidies.[39] Some ten percent of the largest farm groups received 72% of USDA crop subsidies.

More worrisome was the fact that the US Government itself admitted in published reports that its statutory oversight in the health and safety of the nation's meatpacking and processing industry was worse than inadequate. In January 2006, the USDA issued the following report, apparently only in required response to a lone Senator who asked:

> The Grain Inspection, Packers and Stockyards Administration has not established an adequate control structure and environment that allows the agency to oversee and manage its investigative activities for the Packers and Stockyards Programs (P&SP).... P&SP's tracking system could not be relied upon, competition and complex investigations were not being performed, and timely action was not being taken on issues that impact day-to-day activities. These material weaknesses should be reported in the agency's next FMFIA report because they represent essential activities for administering and enforcing the Packers and Stockyards Act of 1921 (Act). The Act prohibits unfair, unjustly discriminatory, and deceptive acts and practices, including certain anti-competitive practices. We also found that the agency has not taken sufficient actions to strengthen operations in response to findings previously reported by the Office of Inspector General (OIG) in February 1997 and the Government Accountability Office (GAO) in September 2000. Our current work was initiated in response to concerns raised by a US Senator in April 2005.[40]

The last statement implied they would not have undertaken such an inquiry on their own.

It was no accident. The powerful Washington lobbyists of agribusiness drafted the Farm Bills that dispersed the funds, and influenced which policies got enforced, as well as the appointment of agribusiness-friendly bureaucrats and officials to enforce them. The 1921 Packers and Stockyards Act had become an empty construct, honored in its breach.

The now powerful forces of the agribusiness lobby scored a major victory in 1996 with passage of the new Farm Bill by the US Congress. US farm policy from 1933, as explicitly stated in the Agricultural Adjustment Act of 1938, during the Great Depression, granted authority to the Secretary of Agriculture to attempt to balance demand and supply, by idling land, implementing commodity storage programs, establishing marketing quotas for some crops and to encouraging exports of commodities including food relief programs and sales of farm commodities for soft currencies. However, after 1996, the Secretary's authorities were suspended, if not repealed, in the 1996 and 2002 farm bills.

Before 1996, sharp price swings were moderated through the use of storage programs and land idling. The costs for the stabilization were relatively modest compared with the costs incurred after 1997. The 1996 farm bill, enacted during a brief period of economic euphoria in 1996, temporarily stripped the Secretary of Agriculture of all authority to manage inventories and set the stage for all-out production of the major program crops. That authority to idle resources, which every other CEO has authority to do when inventories become excessive, was swept away despite overwhelming evidence that agriculture's capacity to produce has consistently exceeded the capacity of markets to absorb the production without resorting to unacceptably low prices. With the transition away from government programs, it was expected that market forces would appropriately throttle resource use in agriculture. The results were a huge boon for agribusiness in their pursuit of ever-larger land at a cheap price. For the family farmer, the price was staggering.

As a report done by Iowa State University concluded:

> Prices declined because the 1996 farm bill no longer authorized the government to idle land to balance demand and supply. Production decisions were left to the market.... When no land is idled, production increases, crop prices fall, and land values come under pressure until there is less profitability for crop production on the least productive land. The market squeezes out the thinner soils and steeper slopes, the higher per-unit cost of production areas. This land then transitions ... to another crop or to grazing land.[41]

Few Americans had the slightest idea of what was going on. By the mid-decade of the new century, however, the general level of public health, the epidemic-scale incidence of obesity, allergies, diseases once rare in the general population such as salmonella poisoning, e-coli, all were becoming every day events.

The stage was set by the end of the 1990's for what Ray Goldberg termed a transformation that he described as "changing our global economy and society more dramatically than any other single event in the history of mankind."[42]

By 1998, Goldberg was 77 years old and extremely active, sitting on the boards of numerous large agribusiness companies such as ADM and Smithfield Foods and advising the World Bank on agribusiness for the developing world. That year, he organized a new university-wide research group at Harvard to examine how the genetic revolution would affect the global food system.

The creator of agribusiness was integrating the gene revolution into the agribusiness revolution as the next phase. He mapped out the transformation of world food consolidation thirty years into the future.

His study calculated that "the traditional agribusiness system, without the pharmaceutical, health and life science segments will be an $8 trillion global industry by 2028. The farming sector value added," he went on, "will have shrunk from 32% in 1950 to 10% .... Whereas food processing and distribution accounted for half of 1950's value added, it will account for over 80% in 2028."[43] For Goldberg, the farmer would become a tiny player in the giant global chain.

Goldberg calculated the addition of entire new sectors created by the latest developments in genetic engineering, including GMO creation of pharmaceutical drugs from genetically-engineered plants, which he called "the agri-ceutical system." He declared, "The addition of life science (biotechnology-ed.) participants in the new agri-ceutical system will increase total value added in 2028 to over $15 trillion and the farmers' share will shrink even further to 7%." He proclaimed, enthusiastically, "the genetic revolution is leading to an industrial convergence of food, health, medicine, fiber and energy businesses."[44]

He might have added that all this was virtually without government regulation or scientific supervision by neutral scientific research organizations. How the gene revolution evolved, would again find the Rockefeller Foundation in a central role. From Green

Revolution to Gene Revolution, the foundation was in the center of developing the strategy and means for transforming how the planet fed itself, or didn't feed itself.

Notes

1. UN Food and Agriculture Organization, *Mobilizing Science for Global Food Security*, Fourth External Review of CIMMYT, (Consultative Group on International Agricultural Research—CGIAR, Rome)—SDR/TAC:IAR/97/9. Also CGIAR, *The Origins of the CGIAR*, http://www.cgiar.org/who/history/origins.html, details the role of the Rockefeller Foundation in creation of both CIMMYT and later CGIAR as the larger global agriculture research body to advance the Rockefeller Foundation's growing agribusiness agenda. See also Robert Anderson, *American Foundations, the Green Revolution and the CGIAR: Intentions, Implementation and Contingencies*, Simon Fraser University, November 2003, http://les.man.ac.uk/government/publications/working_papers_docs/Globalisation/Foundations%20papers%20Anderson.pdf. One of the most detailed critiques of Rockefeller's Green Revolution is made in Harry Cleaver, *The Contradictions of the Green Revolution*, http://www.eco.utexas.edu/facstaff/Cleaver/cleavercontradictions.pdf.

2. Harry Cleaver, *op. cit.*, p. 3.

3. CGIAR, *The Origins of the CGIAR*, in http://www.cgiar.org/who/history/origins.html.

4. *Ibid.* For background on the very influential Rockefeller friend, Maurice Strong, see Elaine Dewar, *Cloak of Green*, Lorimar & Co., Toronto, 1995, p. 254, and Henry Lamb, *Maurice Strong: The New Guy in Your Future!*, January 1997, http://www.sovereignty.net/p/sd/strong.html#3.

5. Harry Cleaver, *op. cit.*, p. 5. A. T. Mosher, *Getting Agriculture Moving*, ADC, New York, 1966, p. 34.

6. *Ibid.*, P. 11. Also, "Who's for DDT?", *Time*, 22 November 1971.

7. A. Parsons, "Philippines: Rebellious Little Brother", *Pacific Research and World Empire Telegram*, January 1971.

8. Jeroen van Wijk, "Hybrids Bred for Superior Yields or for Control?", *Biotechnology and Development Monitor*, 1994, No. 19, pp. 3-5.

9. Harry Cleaver, *op.cit.*, p. 9.

10. *Ibid.*, p. 9.

11. Research Unit for Political Economy (R.U.P.E.), "Economics and Politics of the World Social Forum, Appendix I: Ford Foundation—A Case Study of the Aims of Foreign Funding", *Aspects of India's Economy* September 2003. For background on the postwar close ties between the Ford Foundation and the CIA during the 1950's and 1960's see James Petras, "The Ford Foundation and the CIA: A Documented Case of Philanthropic Collaboration with the Secret Police", *Rebelion*, 15 December 2001, http://www.rebelion.org/petras/english/ford010102.htm.

12. Debashis Mandal and S. K. Ghosh, "Precision Farming—the Emerging Concept of Agriculture for Today and Tomorrow", *Current Science*, 25 December 2000. The authors, authorities on Indian agriculture summarize the impact of the Green Revolution in India: "The green revolution has not only increased productivity, but it has also several negative ecological consequences such as depletion of lands, decline in soil fertility, soil salinization, soil erosion, deterioration of environment, health hazards, poor sustainability of agricultural lands and degradation of biodiversity. Indiscriminate use of pesticides, irrigation and imbalanced fertilization has threatened sustainability."

13. John H. Davis, *Harvard Business Review*, 1956, cited in Geoffrey Lawrence, "Agribusiness", *Capitalism and the Countryside*, Pluto Press, Sydney, 1987. See also Harvard Business School, *The Evolution of an Industry and a Seminar: Agribusiness Seminar*, http://www.exed.hbs.edu/programs/agb/seminar.html.

14. Martin Kohli, "Leontief and the U.S. Bureau of Labor Statistics, 1941-54: Developing a Framework for Measurement," *History of Political Economy*, Vol. 33, Annual Supplement, 2001, pp. 190-191.

15. John D. Rockefeller III, *The Second American Revolution*, Harper & Row, New York, 1973, p. 108.

16. Current Biography, 1967, W. Leontief and Ray Goldberg, "The Evolution of Agribusiness", *Harvard Business School Executive Education Faculty Interviews*, http://www.exed.hbs.edu/faculty/rgoldberg.html. W. Leontief, Studies in the Structure of the American Economy, International Science Press Inc., White Plains, New York, 1953. In its 1956 Annual Report, the Ford Foundation noted the following grant: "Harvard Economic Research Project". In addition to these overall programs, a grant of $240,000 was made to support the activities of the Harvard Economic Research Project over a six-year period. This center, under the direction of Professor Wassily Leontief, was engaged in a series of quantitative studies of the structure of the American economy, focusing mainly on inter-industry relationships and the interconnections between industry and other sectors of the economy. Equal support was contributed by the Rockefeller Foundation, Ford Foundation, *Annual Report*, New York, 1956. A fascinating and controversial report of the implementation of the Leontief Harvard Economic Research Project on the Structure of the American Economy is a document titled, *Silent Weapons for Quiet Wars*. Its authorship is disputed, with attribution to Hartford Van Dyke and to William Cooper, and much speculation exists as to whether it is fact or fiction. The discussion in the report of aspects of the Leontief research, its Rockefeller funding and how it was linked actively with the work of Ray Goldberg and John H. Davis in creating the model of corporate agribusiness is too incisive to dismiss the full report completely. The document for this reason alone is worth reading. http://www.universalway.org/Foreign/silentweapons.html.

17. Roert M Aduddell, and Louis P. Cain, "Public Policy Toward The Greatest Trust in the World", *Business History Review,* Summer 1981, Harvard College, Cambridge, p. 217.

18. *Ibid.,* p. 218.

19. James MacDonald et al., *Growing Farm Size and the Distribution of Farm Payments,* United States Departent of Agriculture, Economic Research Service, Economic Brief No. 6, Washington, D.C. March, 2006, p. 2.

20. The Humane Farming Association, *Factory Farming: The True Costs,* San Rafael California, 31 July 2005, http://www.hfa.org.

21. *Ibid.*

22. Tom Harkin, *Economic Concentration and Structural Change in the Food and Agriculture Sector,* Prepared by the Democratic Staff of the Committee on Agriculture, Nutrition, and Forestry United States Senate, 29 October 2004, p.6.

23. *Ibid.,* pp. 5-6. Also Mark Spitzer, "Industrial Agriculture and Corporate Power", *Global Pesticide Campaigner,* August 2003, http://www.panna.org/iacp.

24. James MacDonald, et al., *op.cit.,* pp.1-4.

25. The Humane Farming Association, *op. cit.*

26. The Economist, "Growing Pains," *The Economist,* US Edition, 25 March 2000.

27. OMB Watch, *OMB Waters Down Standards on Factory-Farm Runoff,* 28 May 2003, http://www.ombwatch.org/article/articleview/1540. See also Natural Resources Defense Council (NRDC), *Facts about Pollution from Livestock Farms,* Washington, D.C., 15 July 2005.

28. OMB Watch, *op. cit.*

29. NRDC, *op. cit.*

30. NRDC, *op. cit.*

31. *Ibid.*

32. OMB Watch, *op. cit.*

33. The Humane Farming Association, *op. cit.*

34. NRDC, *op. cit.*

35. The Humane Farming Association, *op. cit.* See also, Brian DeVore, "Greasing the Way for Factory Bacon, Corporate hog operations—and their lagoons— Threaten the Financial and Physical health of Family Farms", *Sustainable Farming Connection,* http://www.ibiblio.org/farming-connection.

36. Tom Harkin, *op. cit.,* pp. 6-7.

37. Ray Goldberg, *The Genetic Revolution: Transforming our Industry, Its Institutions, and Its Functions,* address to The International Food and Agribusiness

Management Association (IAMA), Chicago, 26 June 2000, pp. 1-2. Goldberg founded and headed the IAMA as well as holding seats on the boards of agribusiness giants Archer Daniels Midland, Smithfield Foods and DuPont Pioneer Hi-Bred. He practiced what he preached.

38. Col. Eddie Coleman, US Army, *Agribusiness Group Paper* National Defense University, 2003, in http://www.ndu.edu/icaf/industry/IS2003/papers/2003%20Agribusiness.htm#.

39. Tom Harkin, *op. cit.*

40. U.S. Department of Agriculture, Office of Inspector General, Northeast Region, *Grain Inspection, Packers and Stockyards Administration's Management and Oversight of the Packers and Stockyards Programs*, Report No. 30601-01-Hy, Washington D.C., January 2006, p. 3.

41. Leopold Center for Sustainable Agriculture Iowa State University, *Toward a Global Food and Agriculture Policy*, January 2005, http://www.leopold.iastate.edu/pubs/staff/policy/globalag.pdf.

42. Ray Goldberg, *The Genetic Revolution, op.cit.*, p. 1.

43. *Ibid.* p. 2. Also see PR Newswire, *Agriceuticals: The Most Important Economic Event in our Lifetime, Says Harvard Professor Dr. Ray Goldberg*, 8 December 1999.

44. *Ibid.* p. 2.

CHAPTER 8
Food is Power…

"Food is power! We use it to change behavior. Some
may call that bribery. We do not apologize."

Catherine Bertini, Executive Director,
United Nations World Food Program,
former US Assistant Secretary of Agriculture [1]

Capturing the Golden Rice Bowl

In 1985, the Rockefeller Foundation initiated the first large-scale
research into the possibility of genetically engineering plants for
commercial use. At the time they termed it a "major, long-term
commitment to plant genetic engineering."[2]

Rockefeller Foundation funds provided the essential catalyst for
the worldwide scientific research and development which would
lead to the creation of genetically modified plants, the Gene
Revolution. Over the following two decades, the Rockefellers would
spend well over $100 million of foundation monies directly, and
several hundred million indirectly, to catalyze and propagate
research on the development of genetic engineering and its application to transform world food production.[3] Clearly, it was a very
big issue in their strategic plans.

In 1982, a group of hand-picked advisers from the Foundation
urged its management to devote future resources to the application
of molecular biology for plant breeding. In December 1984, the
Trustees of the Rockefeller Foundation approved what was seen at
the time as a 10-15 year program to apply new molecular-biological

techniques to the breeding of rice, the dietary staple of a majority of the planet's population.

1984 was the year Ronald Reagan was re-elected to a second term with what he saw as a strong popular mandate to press ahead with his New Right economic agenda of privatization and deregulation, along the lines that had been spelled out by John D. Rockefeller and others more than a decade earlier. American agribusiness had reached a major threshold in terms of its ability to influence USDA agricultural policy and, by extension, the world food market. The time was propitious to initiate a dramatic shift in the future control of the world food supply.

The "New Eugenics": *Reductio ad Absurdum...*

The genetic engineering initiative of the Rockefeller Foundation was no spur of the moment decision. It was the culmination of the research it had funded since the 1930's. During the late 1930's, as the foundation was still deeply involved in funding eugenics in the Third Reich, it began to recruit chemists and physicists to foster the invention of a new science discipline, which it named molecular biology to differentiate it from classical biology. The foundation developed molecular biology as a discipline partly to deflect and blunt growing social criticism of its racist eugenics. Nazi Germany had given eugenics a "bad name."

The Rockefeller Foundation's President during the 1930's, Warren Weaver, was a physicist. He and Max Mason headed the foundation's new biology program. Their largesse in giving funds to scientific research projects gained the foundation enormous influence over the direction of science during the Great Depression by the mere fact they had funds to dispense to leading scientific researchers at a time of acute scarcity. From 1932 to 1957, the Rockefeller Foundation had handed out an impressive $90 million in grants to support the creation of the new field of molecular biology.[4] Molecular biology and the attendant work with genes was a Rockefeller Foundation creation in every sense of the word.

Borrowing generously from their work in race eugenics, the foundation scientists developed the idea of molecular biology from the fundamental assumption that almost all human problems could be "solved" by genetic and chemical manipulations. In the 1938 Annual Report of the Rockefeller Foundation, Weaver first coined the term, "molecular biology" to describe their support for research to apply techniques of symbolic logic and other scientific disciplines to make biology "more scientific." The idea had been promoted during the 1920's by Rockefeller Institute for Medical Research biologist Jacques Loeb, who concluded from his experiments, that echinoderm larvae could be chemically stimulated to develop in the absence of fertilization, and that science would eventually come to control the fundamental processes of biology. The people in and around the Rockefeller institutions saw it as the ultimate means of social control and social engineering, eugenics.[5]

It seemed clear in 1932, when the Rockefeller Foundation launched its quarter-century program in that area, that the biological and medical sciences were ready "for a friendly invasion by the physical sciences". According to Warren Weaver:

> [T]he tools are now available for discovering, on the most disciplined and precise level of molecular actions, how man's central nervous system really operates, how he thinks, learns, remembers, and forgets .... Apart from the fascination of gaining some knowledge of the nature of the mind-brain-body relationship, the practical values in such studies are potentially enormous. Only thus may we gain information about our behavior of the sort that can lead to wise and beneficial control.[6]

During World War II, Weaver and the Rockefeller Foundation were in the center of all international research in molecular biology. Three Rockefeller Institute (today Rockefeller University-w.e.) scientists Avery, MacLeod and McCarty identified what appeared to be the transmission of a gene from one bacterial cell to another. Their colleague, later prominent researcher at the Rockefeller University, geneticist Theodore Dobzhansky, noted at the time with great excitement, "we are dealing with authentic cases of inductions of specific mutations by specific treatments—a feat which geneticists have vainly tried to accomplish in higher organisms." Already in 1941, Rockefeller scientists were laying the foundations for their later development of genetically modified organisms and the Gene Revolution.[7]

Notably, the Rockefeller-funded genetic scientists in the new field of molecular biology congregated at the same Cold Spring Harbor site of the Eugenics Records Office, financed by Carnegie and Rockefeller foundations, to hold major scientific symposiums on the "genetics of micro-organisms" beginning in 1946 just after the war's end.[8]

Reducing Life
Risks entailed weren't interesting to the Rockefeller group. Their methodology went back to what was termed "reductionism" by René Descartes, and to the method of Charles Darwin, namely that living creatures were machines whose only goal was genetic replication—a matter of chemistry and statistics. The Rockefeller methodology was an extension of the belief that a complex life form cold be reduced to a basic building bloc or "elementary seed," from which all traits of the life form could be deduced. It was of little interest to Weaver and others at the Rockefeller Foundation that scientific reductionism had been thoroughly refuted. "Who pays the Piper picks the tune." They had a social agenda and their reductionist genetics supported that agenda.

One scientist critical of the risks of GMO research, Prof. Philip Regal, who organized the first meeting between leading university ecologists and molecular biologists, genetic engineers in industry, and representatives from government agencies, at the Cold Spring Harbor Banbury Center in August 1984, defined the flaw of the molecular biologists' reductionism:

> In the case of DNA, this molecule is stable in a test tube. But it is not stable in populations of reproducing organisms. One cannot reduce the behavior of DNA in living organisms to its chemical properties in a test tube! In living systems, DNA is modified, or "destabilized" if one prefers, at a minimum by mutation, gene flow, recombination, and natural selection. This would make it extremely difficult or even impossible to have a true genetic engineering, in the sense of which it had been spoken. Many molecular biologists certainly "knew" facts

about mutation and natural selection as abstract facts, but they were
not a working part of their professional consciousness.[9]

Once they had made the idea popular in US science that organisms were reduced to genes, they could conclude that organisms
had no inherent nature. Anything was "fair game." But nature was
far more complex than a digital computer. In one example pointed
to by biologists, whereas a given DNA molecule would be stable
in a test tube, it became highly unstable in living organisms, inter-
acting in extremely non-linear and complex ways. Life was not a
binary computer program. It was marvellously non-linear and
complex as traditional biologists had attested for centuries.[10]

The Rockefeller Foundation's molecular biology and their genetics work was consciously based on that fundamental scientific error,
reductionism. Their scientists used the term "genetic programming"
as a metaphor for what happens in a computer, but no scientist was
able to generate an organism from a genetic program. As one British
biologist, Professor Brian Goodwin, pointed out, "You need to know
more than gene products in order to explain the emergence of shape
and form in organisms."[11]

Such details were of no interest to the Rockefeller eugenicists,
who were masquerading in the 1980's as geneticists. More likely
than not, many of the younger generation of biologists and scientists receiving Rockefeller research grants were blissfully unaware
that eugenics and genetics were in any way related. They simply
scrambled for scarce research dollars, and the dollars all too often
had the name and strings of the Rockefeller Foundation attached.

The foundation's research goal was to find ways to reduce the
infinite complexities of life to simple, deterministic and predictive
models. Warren Weaver was intent on using science, bad science if
need be, to shape the world into the Rockefeller model. The promoters of the new molecular biology at the foundation were determined to map the structure of the gene, and to use that information,
as Philip Regal described it, "to correct social and moral problems
including crime, poverty, hunger and political instability."[12] Just
how they would correct such social problems would be kept under
wraps for decades. Regal described the Rockefeller vision:

From the perspective of a theory reductionist, it was logical that social problems would reduce to simple biological problems that could be corrected through chemical manipulations of soils, brains, and genes. Thus the Rockefeller Foundation made a major commitment to using its connections and resources to promote a philosophy of eugenics.

The Rockefeller Foundation used its funds and considerable social, political, and economic connections to promote the idea that society should wait for scientific inventions to solve its problems, and that tampering with the economic and political systems would not be necessary. Patience, and more investment in reductionist research would bring trouble-free solutions to social and economic problems.

Mason and Weaver helped create a network of what would one day be called molecular biologists, that had little traditional knowledge of living organisms and of communities of organisms. It shared a faith in theory reductionism and in determinism. It shared utopian ideals. It learned to use optimistic terms of discourse that brought grants and status. The project was in the general spirit of Bacon's New Atlantis and Enlightenment visions of a trouble-free society based on mastery of nature's laws and scientific/technological progress.[13]

During the 1970's, molecular biologists in the United States intensely debated the issue of whether recombinant DNA research, later referred to as genetic engineering, should at all proceed, or whether, owing to the incalculability of the possible dangers to life on the planet and the risk of an ecological accident, research should be voluntarily ceased in the interest of mankind. By 1973, the essential techniques of genetic engineering had been developed in the laboratory.[14]

One biologist, Dr. Robert Mann, a retired Senior Lecturer of the University of Auckland, emphasised that there was indeed a problem with how Rockefeller reductionist simplification ignored possible social risks: "Attempts at risk analysis for Genetic Engineering are, obviously, doomed to be even more misleading," Mann noted:

> The system of a living cell, even if no viruses or foreign plasmids (let alone prions) are tossed in, is incomparably more complex than a nuclear reactor. There is no prospect of imagining most of the ways it can go badly wrong.... Many gene-splicings come to naught; some others may yield only the desired outcome; but the few major

mishaps, as with nuclear power, dominate the assessment so as to rule out this approach to science and life.[15]

Mann sounded the alarm: one of the countless scientific warnings buried by the powerful agribusiness propaganda machine that stood with the Rockefeller Foundation behind genetically engineered organisms.[16]

"Among the biological materials used for GE," Prof. Abigail Salyers warned in the prestigious *Microbiological Review:*

> [A]re small pieces of DNA called plasmids, depicted ... as simple predictable carriers of engineered genes. According to conventional wisdom, a plasmid used to introduce a gene into a genetically engineered micro-organism can be rendered non-transmissible ... [on the contrary] there is no such thing as a "safe" plasmid ... a riddle we may have to answer in order to survive: what can be done to slow or stop the transfer of antibiotic resistance genes. But the gene jockeys claim they can, Godlike, foresee the evolutionary results of their artificial transposings of human genes into sheep, bovine genes into tomatoes, etc.[17]

The heart of genetic engineering of plants, unlike the long-standing methods of creating plant hybrids by cross-breeding two varieties of the same plant to produce a new variety with specific traits, involved introducing foreign DNA into a given plant. The combining of genes from different organisms was termed recombinant DNA or rDNA. An example was the creation of GE sweet corn or Bt sweet corn. It was made by inserting a gene from a soil bacterium, Bacillus thuringiensis, or Bt, into the genome of a corn variety to protect it from the European Corn Borer pest. In 1961, Bt had been registered as a pesticide. Its ability to combat specific insects was questionable however. One 1999 scientific report warned:

> Evolution of resistance by pests is the most serious threat to the continued efficacy of Bt toxins .... With millions of hectares of Bt toxin-producing transgenic plants grown yearly, other pests are likely to evolve resistance uickly unless effective countermeasures are designed and implemented soon.[18]

Gene transformation usually required a tissue culture or regeneration of an intact plant from a single cell that had been treated with hormones or antibiotics and forced to undergo abnormal development. In order to implant a foreign gene into a plant cell, in addition to a genetically engineered bacteria (Agrobacterium tumefaciens), a "Taxi" or "Gene cannon," a method known also as biolistics, short for bio-ballistics. The gene cannon had been developed in 1987 at Cornell University by John Sanford. Unlike the creation of plant or animal hybrids, genetic engineering of plants bypassed sexual reproduction entirely, and hence was not limited by their species barriers, so that the natural species barriers could be "jumped."[19]

Biologist Dr. Mae-Wan Ho, head of the London Institute of Science in Society, stressed that "entirely new genes and combinations of genes are made in the laboratory and inserted into the genomes of organisms to make genetically modified organisms. Contrary to what you are told by pro-GMO scientists," she went on to say that "the process is not at all precise. It is uncontrollable and unreliable, and typically ends up damaging and scrambling the host genome, with entirely unpredictable consequences."[20]

Neither the Rockefeller Foundation nor the scientists it funded, nor the GMO agribusiness they worked with, had any apparent interest in examining such risks. It was self evident they would have the world believe that risks were minimal.[21]

The first genes had been spliced in 1973 and the recombinant gene technique spread widely among research labs, amid heated debate about the potential risks of misuse of the new technology. There was intense scientific concern about the risk of a so-called "Andromeda Strain" scenario of an escaping mutating species. The term was drawn from science-fiction writer Michael Crichton's 1968 novel, *The Andromeda Strain*, about a deadly disease which causes rapid, fatal clotting of the blood, and threatens all life on Earth.

In 1984, no serious scientific consensus existed within America's research laboratories on the dangers of releasing genetically modified plants into a natural environment. Yet, despite the fact that very significant doubts persisted, the Rockefeller Foundation made the decision to devote major global funding to the genetic modification process.

One very important effect of the Reagan deregulation revolution on the field of molecular biology in the 1980's was that decisions on safety or risks made until then by relatively independent Government agencies were increasingly put in the hands of private companies who saw major gains in advancing the emerging potentials of biotechnology. Rockefeller's planners had little trouble interesting major companies to join in with them in the new experiments in genetic engineering.

Mapping the Rice Genome

In 1984, the Foundation decided to launch its comprehensive program to map the rice genome using new molecular-based techniques and advances in computing power. At the time, there existed no experimental evidence to justify that decision.

Publicly, they announced that their huge research effort was an attempt to deal with world hunger in coming decades, as projected world population growth should add billions of new hungry mouths to be fed. The research monies were channelled through a new entity they created, the International Program on Rice Biotechnology (IPRB), into some of the world's leading research labs. Over the next 17 years, the foundation spent an impressive $105 millions of its own money in developing and spreading genetically modified rice around the world. Furthermore, by 1989 it was spending an additional $54 million a year—amounting to more than $540 million over the following decade—on "training and capacity building" to disseminate the new developments in rice genetic modification. The seeds of the Gene Revolution were being planted very carefully.

"Golden Rice" and Black Lies

The decision to develop a genetically-modified variety of rice was a master stroke of public relations on the part of the Rockefeller Foundation and its supporters within the scientific and political establishment.

Initially, the Foundation funded 46 science labs across the industrialized world. By 1987, they were spending more than $5 million a year on the rice gene project, mapping the rice genome. Among the recipients of Rockefeller largesse were the Swiss Federal Institute of Technology in Zurich and the Center for Applied Biosciences at Freiburg University in Germany.

The grants also went to train a network of international scientists in mastering the worldview of the Rockefeller Foundation's worldview, regarding the role of genetic engineering of plants and the future of humankind. The foundation financed the training of hundreds of graduate students and postdoctoral fellows around the world to create the scientific infrastructure for the later commercial proliferation of genetically modified organisms.

They developed an elite fraternity and cultivated, according to some participants, a strong sense of belonging. The top five scientific researchers at the important Rockefeller-funded Philippine International Rice Research Institute (IRRI) were all Rockefeller-funded doctors. "Without the support of the Rockefeller Foundation it would have been almost impossible for us to build this capability," remarked the IRRI's Deputy Director for Research.[22]

Soon after the program started, the Rockefeller's International Program on Rice Biotechnology (IPRB) decided to concentrate efforts on the creation of a variety of rice which allegedly would address Vitamin A deficiency in undernourished children in the developing world. It was a brilliant propaganda ploy. It helped to create a public perception that genetic scientists were diligently working to solve problems of world hunger and malnourishment. The only problem was that it was a deliberate deception.

The choice of rice to begin the Rockefeller's gene revolution was a careful one. As one researcher pointed out, rice is the staple food for more than 2.4 billion people. It had been domesticated and developed by local farmers over a period of at least 12,000 years, and has grown in a wide variety of different environments.[23]

Rice was synonymous with food security for most of Asia, where over 90% of the global rice harvest was produced, primarily by China and India, and where it made up 80% of people's daily calories. Rice was also a staple in West Africa, the Caribbean and tropical regions of Latin America. Rice farmers had developed varieties of rice to withstand droughts, resist pests, and grow in every climate imaginable, all without the help of biotechnology. They had created an incredible biological diversity with over 140,000 varieties.[24]

The Rockefeller Foundation had its eyes on Asia's rice bowl well before the 1984 IPRB project on rice. A prime target of the foundation's Green Revolution had been Asian rice production. The Green Revolution process had significantly destroyed the rich rice diversity over a period of thirty years, with the so-called High-Yielding Varieties. This drew Asia's peasantry into the vortex of the world trade system and the global market for fertilizer, high-yielding seeds, pesticides, mechanisation, irrigation, credit and marketing schemes packaged for them by Western agribusiness.

The core driver of that earlier rice revolution had been the Philippines-based Rockefeller Foundation-created International Rice Research Institute (IRRI). It was not surprising then, that the IRRI, with a gene bank containing more than one-fifth of the world's rice varieties, became the prime vehicle to proliferate the Rockefeller Foundation's new gene revolution in rice. They banked every significant rice variety known.

IRRI had been used by the backers of the Green Revolution to gather control of the irreplaceable seed treasure of Asia's rice varieties, under the ruse that they would thereby be "protected."

The IRRI was put under the umbrella of the Consultative Group of International Agricultural Research, CGIAR, after its creation in 1960 by the Rockefeller and Ford Foundations during the Green Revolution in Asia. CGIAR was the same agency which also controlled the pre-war Iraqi seed bank. CGIAR operated out of the World Bank headquarters in Washington, also with Rockefeller Foundation funding.[25]

In this manner, the World Bank, whose political agenda was defined by Washington policy, held the key to Asia's rice seed bank. Over three-quarters of the American rice genetic makeup or germplasm came originally from the IRRI seed bank. That rice was then pressed on Asian countries by the US Government demanding that Asian countries remove "unfair trade barriers" to US rice imports.

IRRI then became the mechanism for allowing major international agribusiness giants like Syngenta or Monsanto to illegally take the seeds from the IRRI seed bank, initially held in trust for the native farmers of the region.

The seeds, once in the labs of Monsanto or the other biotech giants, would be modified genetically, and patented as exclusive intellectual property of the biotech company. The World Trade Organization, created in 1994 out of the GATT Uruguay Round, introduced a radical new Agreement on Trade-Related Aspects of Intellectual Property Rights (TRIPS) permitting multinationals to patent plant and other life forms for the first time.

In 1993, a Convention on Biological Diversity under the UN was agreed upon to control the theft of such seed resources of the developing world. Washington, however, made a tiny alteration in the original text. It demanded that all the genetic resources held by the CGIAR system (of which IRRI is part) remain outside the rules. That affected half a million seed accessions, or 40% of the world's unique food crop germ-plasm held in gene banks. It meant that agribusiness companies were still free to take and patent them.[26]

Using the IRRI resources as its center, Rockefeller financing for Vitamin-A-enhanced rice became the prime focus of the IPRB research by the beginning of the 1990's. Their grants financed major work in the area by, among others, the Swiss Federal Institute of Technology in Zurich.

The Foundation's propagandists argued that lack of Vitamin A was a major cause of blindness and death in newborn infants in developing countries. UN statistics indicated that perhaps 100 to 140 million children worldwide had some form of Vitamin A deficiency, and among them 250,000 to 500,000 went blind. It was a human interest story of prime emotional attraction to promote acceptance of the controversial new genetically modified plants and crops. Golden Rice became the symbol, the rallying flag, and the demonstration of the promise of genetic engineering, even though the promise was based on black lies and deliberate deception.

The introduction of genetically modified rice would open the prospect of directly controlling the rice seeds, the basic food staple of 2.4 billion people. Prior to the gene revolution, rice had been ignored by the multinational agribusiness seed companies. That in part owed to the low income of rice regions and their peasants, and in part to the fact that rice had proven extremely difficult to hybridize. Farmer-saved seed accounted for more than 80% of Asian rice seed.

In their effort to take over this huge rice market with genetically modified seeds, the foundation and its agribusiness collaborators left nothing to chance or to the vagaries of the free markets. In 1991, the Rockefeller Foundation and the Rockefeller Brothers Fund created a new organization, the International Service for the Acquisition of Agri-biotech Applications (ISAAA), headed by the Rockefeller Foundation's Mexican Green Revolutionary and head of the CIMMYT or International Center for Wheat and Maize Improvement, Dr. Clive James.[27]

The purpose of ISAAA was, in their own words, to "contribute to poverty alleviation in developing countries by increasing crop productivity and incomes, particularly among resource-poor farmers, and to bring about more sustainable agricultural development in a safer global environment."[28] The only hitch in the deal was that this formidable task, according to their framework, could only be done by the use of biotechnology.

The ISAAA was merely a platform to proliferate genetically engineered plants in target developing sector countries. It had been created and put into motion almost a full decade before it was clear that the Rockefeller Foundation's Golden Rice development was even feasible. It was from its outset, intent on proliferating gene plants in developing countries.

But the Foundation was not alone in backing ISAAA. The ISAAA was also backed financially by biotech agribusiness corporations such as Monsanto, Novartis (Syngenta), AgrEvo (Aventis Crop Science) and the US State Department's USAID. Their goal was to "create global partnerships" between the agribusiness biotech giants of the industrialized countries (notably, the USA) and the developing countries. To create those partnerships, ISAAA set up technology transfer projects covering the topics of tissue culture, diagnostics or genetic engineering.[29]

Interestingly enough, just as Henry Kissinger compiled a list of 13 "priority" developing countries for US Government depopulation policies in his NSSM 200 strategy document of 1974, the ISAAA also developed a priority target list for the introduction of genetically engineered plants and crops. The list of 12 countries included Indonesia, Malaysia, the Philippines, Thailand, and Vietnam in Asia; Kenya, Egypt, and Zimbabwe in Africa; and Argentina, Brazil, Costa Rica, and Mexico in Latin America. Significantly, half of the ISAAA priority countries overlapped with Kissinger's geopolitical targets of seventeen years prior. Indeed, geopolitics presented certain constants.[30]

By 2000, the Rockefeller Foundation and the Swiss Federal Institute of Technology announced that they had successfully taken two genes from a daffodil, together with a gene of a bacterium, and built it into the rice DNA in order to produce what they called pro-Vitamin A or beta carotene rice.

Because the beta-carotene (or pro-Vitamin A) which produced Vitamin A inside the body colored the rice grain orange, it was dubbed "Golden Rice"—another brilliant marketing stroke as everyone covets gold in whatever form. Now people could ostensibly get their daily bowl of rice and prevent blindness and other manifestations of Vitamin A deficiency in their children at the same time.

Children in Asia and the rest of the world had been receiving Vitamin A from other sources for centuries. The problem was not lack of natural foods containing Vitamin A, but rather not enough access to those other natural sources of Vitamin A.

Indian biodiversity campaigner, Dr. Vandana Shiva, pointed out in a stinging critique of the Rockefeller Foundation Golden Rice promotion that, "the first deficiency of genetic engineering rice to produce Vitamin A is the eclipsing of alternative sources of vitamin A". Per Pinstripe Anderson, head of the International Rice Research Institute, has said that Vitamin A rice is necessary for the poor in Asia because "we cannot reach very many of the malnourished in the world with pills."[31]

Shiva pointed out, "there are many alternatives to pills for Vitamin A supply. Vitamin A is provided by liver, egg yolk, chicken, meat, milk, butter. Beta-carotene, the Vitamin A precursor is provided by dark green leafy vegetables, spinach, carrot, pumpkin, mango . . . ."[32]

Not mentioned in Rockefeller Foundation press releases, doctors and scientists knew that large quantities of Vitamin A could in fact lead to "hypervitaminosis," or Vitamin A toxicity which, in infants, could lead to permanent brain damage and other harmful effects.[33]

Moreover, the quantity of rice which a person would have to consume daily to meet the full quota of Vitamin A was staggering, and not humanly possible. One estimate was that an average Asian would have to eat 9 kilograms of cooked rice daily, just to get the required minimum intake of Vitamin A. A typical daily ration in Asia of 300 grams rice would provide only 8% of his daily requirement.[34]

The Rockefeller Foundation's President Gordon Conway sheepishly responded to these criticisms in a press release: "First it should be stated that we do not consider golden rice to be the solution to the vitamin A deficiency problem. Rather it provides an excellent complement to fruits, vegetables and animal products in diets, and to various fortified foods and vitamin supplements." He added: "I agree with Dr. Shiva that the public relations uses of golden rice have gone too far."[35]

Maybe the "public relations uses" had gone too far, but the campaign to proliferate genetically-modified Golden Rice had obviously not gone far enough for those behind the Rockefeller Foundation's gene revolution.

The Rockefeller Foundation announced in 2000 that it was turning the results of its years of rice research over to the public. In fact, they shrewdly turned it over to the agribusiness biotechnology giants. The UK firm, AstraZeneca, later part of the Swiss Syngenta Company, announced in May 2000 that it had acquired exclusive rights to commercialize Golden Rice.

Golden Rice gave the genetic engineering biotech industry a huge propaganda tool. In 1999, then-President Bill Clinton declared, "If we could get more of this golden rice, which is a genetically

modified strain of rice especially rich in vitamin A, out to the
developing world, it could save 4,000 lives a day, people that are
malnourished and dying."[36] Syngenta and also Monsanto licensed
patents on Golden Rice claiming that they would allow the technology to "be made available free of charge for humanitarian uses
in any developing nation."[37]

The criticism and skepticism about the wisdom of turning our
basic food staples over to the gene doctors and agribusiness giants
grew weaker as the propaganda machine of the Rockefeller
Foundation and the agribusiness lobby went into high gear. One
very prominent medical expert, Dr. Richard Horton, editor of the
British medical journal *The Lancet*, said, "Seeking a technological
food fix for world hunger may be ... the most commercially male-
volent wild goose chase of the new century."[38] Few listened.

An insider in the world of biotechnology, Steven Smith, who
worked on genetic engineering of seeds for the Swiss Syngenta
Seeds, the main holder of the Golden Rice patents, declared shortly
before his death in June 2003, "If anyone tells you that GM is going
to feed the world, tell them that it is not.... To feed the world takes
political and financial will—it's not about production and distribution."[39] The Rockefeller Foundation claim about feeding the
world with genetically modified organisms was just a myth. But it
was a myth in the hands of a powerful mythmaker. The revolution
proceeded.

With an elaborate international structure for proliferating the
seeds of the gene revolution through ISAAA, CGIAR, IRRI and the
direct funding of the Rockefeller Foundation, agribusiness and the
backers of the gene revolution were ready for the next giant step:
the consolidation of global control over humankind's food supply.
For that, a new organization became indispensable. It was called the
World Trade Organization.

Notes

1. Catherine Bertini, U.N. 4th World Conference on Women, Beijng, China, September 1995, cited in *Famous Quotes and Quotations about UN*, http://www.quotes.liberty-tree.ca/quotes_about/un. The much-honored civil servant and winner of World Food Prize in 2003, Bertini, is a former Confidential Assistant to New York State Governor Nelson Rockefeller and a member of David Rockefeller's Trilateral Commission. The World Food Prize, interestingly, was created by former Rockefeller Foundation agronomist, Norman Borlaug, creator of the first Green Revolution, in 1986.

2. Gary H. Toenniessen, *Vitamin A Deficiency and Golden Rice: The Role of the Rockefeller Foundation*, 14 November 2000, http://www.rockfound.org/library/111400ght.pdf, p. 3. Toenniessen, the director of Food Security at the Rockefeller Foundation described his work as follows: "In the early 1980s, advances in plant molecular biology offered the promise of achieving genetic improvements in crops that could not be accomplished with conventional plant breeding. For the most part, however, such advances in crop biotechnology were not being applied to rice or other food crops of primary importance in developing countries. To help make sure the benefits of this powerful new technology would be available to poor farmers and consumers, the Rockefeller Foundation, beginning in 1985, committed roughly half of its agricultural funding to an international program on rice biotechnology. The primary objective of this program was to build rice biotechnology capacity in Asia, and an important part of it was funding the training of Asian scientists at advanced Western laboratories, where they invented techniques and worked on traits important for genetic improvement of rice-skills and knowledge which they then brought back home."

3. J.C. O'Toole, G.H. Toenniessen et al., *The Rockefeller Foundation's International Program on Rice Biotechnology*, Rockefeller Foundation archives, http://www.rockfound.org/library/01rice_bio.pdf.

4. Philip J. Regal, *A Brief History of Biotechnology Risk: The Engineering Ideal in Biology*, Edmonds Institute, 18 July 1999, http://www.cbs.umn.edu/~pregal/GEhistory.htm.

5. Pnina Abir-Am, *The Biotheoretical Gathering, Transdisciplinary Authority and the Incipient Legitimation of Molecular Biology in the 1930s: New Perspectives on the Historical Sociology of Science*. Hist. Sci. 25:1-70, 1987, pp. 18-22, 33.

6. Cited in Robert Bruce Baird, *We Can Change the World*, http://www.government.articlesarchive.net/we-can-change-the-world.html.

7. Joshua Lederberg, "The Impact of Basic Research in Genetic Recombination-A Personal Account", Part I, *Annual Review of Genetics*, Vol. 21, 1987, p. 186.

8. Joshua Lederberg, *ibid.*, Part II, p. 196.

9. Philip J. Regal, *op. cit.*, *The Engineering Ideal in Biology*. See also, Richard Milton, *Shattering the Myths of Darwinism*, http://www.alternativescience.com/shattering-the-myths-of-darwinism.htm.

10. Richard Milton, *op. cit.*

11. David King, "An Interview with Professor Brian Goodwin", *GenEthics News*, March/April 1996, pp. 6-8. Goodwin explains his concerns about genetic or biological reductionism in the interview: "We currently experience crises of health, of the environment, of the community. I think they are all related…Biology contributes to these crises by failing to give us adequate conceptual understanding of life and wholes, of ecosystems, of the biosphere, and it's all because of genetic reductionism…Let me just describe some of the consequences of genetic reductionism. Once you've got organisms reduced to genes, then organisms have no inherent natures. Now, in our theory of evolution, species are natural kinds, they are really like the elements, if you like. I don't mean literally, but they have the same conceptual status, gold has a certain nature. We are arguing that, say, a sea urchin of a particular species has a nature. Human beings have a nature. Now, in Darwinism, they don't have a nature, because they're historical individuals, which arise as a result of accidents. All they have done is pass the survival test. The Darwinian theory makes it legitimate to shunt genes around from any one species to any other species: since species don't have 'natures', we can manipulate them in any way and create new organisms that survive in our culture. So this is why you get people saying that there is really no difference between the creation of transgenic organisms, that is moving genes across species boundaries, and creating new combinations of genes by sexual recombination within species. They say that is no different to what is happening in evolution…Once you scale something up to a particular level you are into a totally different scene. Now, I think that there are the same problems that arise with respect to creation of transgenics, and the reason is because of the utter unpredictability of the consequences of transferring a gene from one species to another. Genes are defined by context. Genes are not stable bits of information that can be shunted around and express themselves independently of context. Every gene depends upon its context. If you change the context, you change the activity of the gene. ..I'm by no means against biotechnology. I just think that it is something that we have to use with enormous caution in its application. We need stringent safety protocols."

12. Philip J.Regal, *op. cit.*

13. Philip Regal, *Metaphysics in Genetic Engineering: 2.2 Utopianism*, paper prepared for International Center for Human and Public Affairs, Buenos Aires, 1996, http://www.psrast.org/pjrbiosafety.htm. Regal adds, "from the perspective of a theory reductionist, it was logical that social problems would reduce to simple biological problems that could be corrected through chemical manipulations of

soils, brains, and genes. Thus the Rockefeller Foundation made a major commitment to using its connections and resources to promote a philosophy of eugenics. The Rockefeller Foundation used its funds and considerable social, political, and economic connections to promote the idea that society should wait for scientific inventions to solve its problems, and that tampering with the economic and political systems would not be necessary. Patience, and more investment in reductionist research would bring trouble-free solutions to social and economic problems.

Mason and Weaver helped create a network of what would one day be called molecular biologists, that had little traditional knowledge of living organisms and of communities of organisms. It shared a faith in theory reductionism and in determinism. It shared utopian ideals. It learned to use optimistic terms of discourse that brought grants and status. The project was in the general spirit of Bacon's New Atlantis and Enlightenment visions of a trouble-free society based on mastery of nature's laws and scientific/technological progress (e.g. Eamon 1994, Mcknight 1992)."

14. Philip J. Regal, *A Brief History of Biotechnology Risk Debates and Policies in the United States*, 18 July 1999, http://www.cbs.umn.edu/~pregal/GEhistory.htm.

15. Dr. Robert Mann, "The Selfish Commercial Gene", *Prast.* http://www.psrast.org/selfshgen.htm. Mann adds the clear warning: "The hazards of GE rival even nuclear war. Biology is so much more complex than technology that we should not pretend we can imagine all the horror scenarios, but it is suspected that some artificial genetic manipulations create the potential to derange the biosphere for longer than any civilisation could survive. If only enthusiasts are consulted in appraisal of GE proposals, such scenarios will not be thought of."

16. Philip J. Regal, *op. cit.*

17. Abigail Salyers, cited in Dr. Robert Mann, *op. cit.*

18. David G. Heckel, et al., "Genetic Mapping of Resistance to Bacillus Thuringiensis Toxins in Diamondback Moth Using Biphasic Linkage Analysis, Proceedings of the National Academy of Sciences, USA", *Agricultural Sciences*, July 1999.

19. Mae-Wan Ho, *FAQ on Genetic Engineering*, Institute of Science in Society, in http://www.i-sis.org.uk/FAQ.php.

20. Mae-Wan Ho, *Puncturing the GM Myths*, http://www.unobeserver.com, 4 August 2004.

21. Philip J. Regal, *op.cit.*

22. Dennis Normile, "Rockefeller to End Network After 15 Years of Success", *Science*, 19 November 1999, pp. 1468–1469, reprinted in ww.gene.ch/genet/2000/Feb/msg00005.html.

23. Gary H. Toenniessen, "Vitamin A Deficiency and Golden Rice: The Role of the Rockefeller Foundation", *The Rockefeller Foundation*, 14 November 2000, in http://www.rockfound.org/library/111400ght.pdf.

24. M. T. Jackson, "Protecting the Heritage of Rice Biodiversity", *GeoJournal* March 1995, pp 267-274. Quoted in K.S. Fisher (editor), "Caring for the Biodiversity of Tropical Rice Ecosystems", *IRRI*, 1996. See also Anna-Rosa Martinez I. Prat, "Genentech Preys on the Paddy Field", *Grain*, June 1998.

25. In its 1998 report, *Shaping CGIAR's Future*, October 26-30, 1998, http://www.worldbank.org/html/cgiar/publications/icw98/icw98sop.pdf, the World Bank authors stated, "World Bank President James D. Wolfensohn formally opened ICW98...Wolfensohn praised the CGIAR's "extraordinary achievements" and recalled that one of his first lessons in development economics was at the hands of his colleagues in the CGIAR. As a board member of the Rockefeller Foundation thirty years ago, he visited CIMMYT in Mexico, where he walked through the fields with local farmers. Evoking this fond memory, Mr. Wolfensohn expressed his "very, very strong and very deep feeling" for the CGIAR. Maurice Strong had worked with David Rockefeller and the family since 1947, and became a Trustee of the Rockefeller Foundation which provided funds for the UN Stockholm Earth Summit in 1972; the latter catalyzed an international movement around the Club of Rome "Limits to Growth" scarce resources report. See Henry Lamb, *Maurice Strong: The New Guy in Your Future!*, http://www.sovereignty.net/p/sd/strong.html, January 1997.

26. The Crucible II Group, *Seeding Solutions: Volume 1: Policy Options for Genetic Resources, Policy primer Major changes in the policy environment*, in http://www.idrc.ca/en/ev-64406-201-1-DO_TOPIC.html.

27. Devlin Kuyek, "ISAAA in Asia: Promoting Corporate Profits in the Name of the Poor", *GRAIN*, October 2000, http://www.grain.org/publications/reports/isaaa.htm.

28. *Ibid.*

29. *Ibid.*

30. *Ibid.*

31. Dr. Vandana Shiva, *Genetically Engineered Vitamin "A" Rice: A Blind Approach to Blindness Prevention*, http://www.biotech-info.net/blind_rice.html, 14 February 2000.

32. *Ibid.*

33. Razak Lajis, "Vitamin A Toxicity", http://www.prn2.usm.my/mainsite/bulletin/sun/1996/sun43.html. Original report cited is from *Australian Adverse Drug Reactions Bulletin*, Vol. 15, No.4, November 1996, which notes, "ADRAC

recently reviewed a report of a child born with microcephaly and dystonia whose mother has inadvertently ingested large quantities of vitamin A during the first four or five weeks of pregnancy. The child subsequently died. While it was not possible to implicate the ingestion of vitamin A as a definite cause of the birth defects in this case, excess amounts of vitamin A are suspected causes of birth defects and its therapeutically used congeners are established causes of birth defects". See also, Marion Nestle, "Genetically Engineered Golden Rice is Unlikely to Overcome Vitamin A Deficiency", Letter to the Editor, *Journal of the American Dietetic Association*, March 2001, pp. 289-290.

34. Benedikt Haerlin, *Opinion Piece about Golden Rice*, archive.greenpeace.org/ geneng/highlights/food/benny.htm. Also, Assisi Foundation, BIOTHAI et al., *Biopiracy, TRIPS and the Patenting of Asia's Rice Bowl*, http://www.poptel.org.uk/ panap/archives/larice.htm, May 1998.

35. Gordon Conway quoted in Paul Brown, "GM Rice Promoters Have Gone too Far", *The Guardian*, 10 February 2001.

36. Bill Clinton, quoted in Paul Brown, *op. cit.*

37. Paul Brown, *op. cit.*

38 Richard Horton quoted in Alex Kirby, "'Mirage' of GM's Golden Promise", *BBC News Online*, 24 September 2003.

39. Alex Kirby, *op. cit.*

PART IV
Unleashing GMO Seeds

CHAPTER 9
A Revolution in World Food Production Begins

Argentina is the First Guinea Pig

By the end of the 1980's, a global network of genetically-trained molecular biologists had developed. A mammoth Rockefeller GMO project was launched. Its chosen location was Argentina, where David Rockefeller and Rockefeller's Chase Manhattan Bank had cultivated close ties to the newly-named President, Carlos Menem. The agricultural land and the population of Argentina were slated to become the first mass testing ground, the first guinea pigs for GMO crops.

Its backers hailed the introduction of GMO agriculture as nothing less than a "Second Green Revolution," a reference to the introduction of modern agriculture production techniques after World War II. In particular, special wheat hybrids and chemical fertilizers were promoted under the rubric that they would increase per hectare crop yields in Mexico, India and other developing lands.

In a short space of just eight years, worldwide acreage planted with GMO crops grew to 167 million acres by 2004, an increase of some 40-fold. That acreage represented an impressive 25% of total land under agricultural cultivation in the world, suggesting GMO crops were well on the way to fully dominating world crop production, at least in basic crops, within a decade or even less.

Over two-thirds of that acreage, or 106 million acres, was planted by the world's leading GMO advocate, the United States. That fact, its proponents argued, proved there was a high degree of confidence on the part of the US Government and consumers, as well as farmers, that GMO crops offered substantial benefits over conventional crops.

By 2004, Argentina was second after the United States in size of acreage planted with GMO crops, with 34 million acres of planting. Far smaller but fast-expanding GMO countries included Brazil, which in early 2005 repealed a law banning planting GMO crops. They argued the crops had already proliferated so widely it was not possible to control the spread. Canada, South Africa and China all had significant GMO crop programs in place by then.

Close behind them and moving fast to catch up were Romania, Bulgaria and Poland, former Soviet Union satellites, rich in agricultural land and loose in regulations. Indonesia, the Philippines, India, Columbia, Honduras and Spain also reported significant GMO plantings. According to data compiled by the Pew Foundation of the United States, many other poorer countries were reported to have been targeted by companies promoting their GMO crops and special herbicide and pesticide chemicals.[1]

According to the Pew study, 85% of farmers planting GMO crops in 2004 were "resource poor". Most were in developing countries, the same countries struggling with IMF reforms and high foreign debts.

No country saw such a radical transformation, and at such an early stage of its fundamental structure of agriculture holdings, as did Argentina. The history of GMO agriculture and the Argentine Soybean revolution was a case study for a nation's systematic loss of food self-sufficiency in the name of "progress."

Up to the beginning of the 1980's, Argentina had been remarkable for the standard of living it provided its population. The agricultural system, partly as a result of the Juan Peron era, was diverse, productive and dominated by small family farms. A typical Argentine farmer in the 1970's would raise a small amount of crops such as vegetables or wheat, husband small poultry, a dairy herd and occasionally beef cattle on a small plot of land, which was held over decades by right of possession. Argentine beef quality was so high in the 1970's that it rivaled that of Texas beef as the world's highest standard of quality. Up to the 1980's, the rich land and farm culture typically produced large surpluses beyond domestic food needs. Significantly, government farm subsidies were non-existent and farmer debts were minimal.

How a Debt Crisis Makes Argentina a Soybean Giant

That all changed with the 1980's Argentina debt crisis. Following the sharp rise in worldwide oil prices during the 1970's, international banks, led by the Rockefeller family bank, Chase Manhattan, Citibank, Chemical Bank, Bank of Boston, and Barclays, among others, sold loans to countries like Argentina on initially very attractive terms. The loans were to finance the import of much-needed oil, among other things. As long as London interest rates remained low, those loans could be serviced from national income. The loans therefore quickly proved to be enormously alluring, and so, the dollar debts rose drastically.

In October 1979, in order to prevent the dollar from collapsing, the US Federal Reserve suddenly raised its major interest rate by some 300%, impacting worldwide interest rates, and above all the floating rate of interest on Argentina's foreign debt.

By 1982, Argentina was caught in a debt trap not unlike that which the British had used in the 1880's to take control of the Suez Canal from Egypt. New York bankers, led by David Rockefeller had learned the lessons of British debt imperialism.[2]

Breaking Argentina's National Will

During its earlier era of Peronism, Argentina had combined a strong and well organized trade union movement, with a central state heavily involved in the economy. Both of these cooperated with select private companies under a regulated model. During the peaceful era of postwar world economic expansion, it had certain features similar to the Scandinavian social democratic model. Furthermore, Peronism, whatever its shortfalls, had created a strong national identity among the Argentine population.

The Peron era came to a bloody end in 1976 with a military coup and regime change backed by Washington. The coup was justified on the argument that it was to counter growing terrorism and communist insurgency in the country. Later investigations revealed that the guerrilla danger from the People's Revolutionary Army (ERP) and the Montoneros had been fabricated by the Argentine military, most of whose leaders had been trained in domestic counter-insurgency techniques by the US Pentagon in the notorious US Army School of the Americas.

The military dictatorship of President Jorge Videla, however, would turn out to be too liberal in its definition of human rights and due process of law. In October 1976, Argentine Foreign Minister, Admiral Cesar Guzzetti met with Secretary of State Henry Kissinger and Vice President Nelson Rockefeller in Washington. The meeting was to discuss the military junta's proposal for massive repression of opposition in the country. According to declassified US State Department documents released only years later, Kissinger and Rockefeller not only indicated their approval, but Rockefeller even suggested specific key individuals in Argentina to be targeted for elimination.[3] At least 15,000 intellectuals, labor leaders and opposition figures disappeared in the so-called "dirty war."

The Rockefeller family played more than an incidental role in the Argentine regime change. A key actor in the junta regime, Economics Minister Martinez de Hoz, had close connections to David Rockefeller's Chase Manhattan Bank and was a personal friend of his. Martinez de Hoz was head of the wealthiest landowning family in Argentina. He introduced radical economic policies designed to favor foreign investment in Argentina. In fact, this economic maneuvering was the very reason behind Rockefeller's secret backing of the Junta in the first place. Large infusions of cash from Rockefeller's bank had privately financed the military's seizure of power.

The Rockefeller brothers regarded Latin America as a de facto private family sphere of influence at least since the 1940's, when David's brother Nelson was running US intelligence in the Americas as President Roosevelt's Coordinator of Inter-American Intelligence Affairs (CIAA). Rockefeller family interests had spread from Venezuelan oil to Brazilian agriculture. Now they had decided that the 1970's debt problems of Argentina offered a unique opportunity to advance family interests there.

While freezing wages, Martinez de Hoz freed domestic wages and prices on the necessities which had been under government price control, including food and fuel, leading to a substantial drop in consumer purchasing power. Import tariffs were slashed, allowing imports to flood the market. The peso-dollar exchange rate was the main nominal anchor of the scheme. Indeed, the budget deficit was reduced from 10.3% of GDP in 1975 to 2.7% in 1979 through expenditure cuts, public-sector price increases, and tax increases, and the inflation rate fell from 335% in 1975, to 87.6% in 1980. However, the real appreciation of the peso, and the resulting capital flight and balance of payments crisis, led to the collapse of the program.[4] Foreign speculative capital was also ushered into the country, and Chase Manhattan and Citibank were the first foreign banks to make their entrance.

Inevitably, there was protest from the strong Peronist union movement against the attack on living standards, which protest the military regime brutally suppressed, along with all other forms of opposition. Clearly satisfied with the new Argentine government, David Rockefeller declared, "I have the impression that finally Argentina has a regime which understands the private enterprise system."[5]

By 1989, following more than a decade of repressive military rule, a new phase in the erosion of Argentine national sovereignty was introduced with the accession of President Carlos Menem, a wealthy playboy later accused of rampant corruption and illegal arms dealing. George Herbert Walker Bush was then in the White House, and received Menem as personal White House guest no less than eight times. His son, Neil Bush, was a guest at Menem's residence in Buenos Aires. Menem, in short, enjoyed the best connections in the North.

With the Argentine military ridden with scandal and with popular discontent growing, New York bankers and Washington power brokers decided it was time to play a new card to continue their economic plunder and corporate takeover of Argentina. Menem was a Peronist only in party name. In fact, he imposed on Argentina an economic shock therapy even more drastic than Margaret Thatcher's British free market revolution of the 1980's. But his Peronist membership allowed him to disarm internal resistance within the party and the unions.

For powerful New York bankers, the key post in the Menem government was the Economics Minister. The new minister was Domingo Cavallo, a disciple of Martinez de Hoz, and a man well-known in New York financial circles. Cavallo got his PhD at David Rockefeller's Harvard University, had briefly served as head of the National Bank, and was openly praised by Rockefeller.[6]

Cavallo was also a close friend and business associate of David Mulford, President George H.W. Bush's key Treasury Official responsible for the restructuring of the Latin American debt under the Brady Plan, and later a member of Credit Suisse First Boston bank. Cavallo was indeed trusted by the "Yankee bankers."[7]

Menem's economic program was written by David Rockefeller's friends in Washington and New York. It gave priority to radical economic liberalization and privatization of the state, and dismantled carefully enacted state regulations in every area from health, to education, to industry. It opened protected markets to foreign imports even further than had been possible under the military junta. The privatization agenda had been demanded by Washington and the IMF—which was acting on Washington's behalf—as a condition for emergency loans to "stabilize" the Peso. At the time, Argentina was suffering from a Weimar-style hyperinflation rate of 200% a month. The Junta had left behind them a wrecked economic and fiscal economy, deeply in debt to foreign banks.

Menem was able to take advantage of the hyperinflation which was engineered during the final years of the Junta, and imposed on the country economic change far more radical than even the military dictatorship had dared. Cavallo dutifully imposed the demanded shocks, and got an immediate $2.4 billion credit, and high praise, from the IMF. A wave of privatizations followed, from the state telecommunications company to the state oil monopoly, and even to Social Security state pensions. Corruption was rampant. Menem's cronies became billionaires at the taxpayer's expense.

In place of state monopolies on industry, giant foreign-owned private monopolies emerged, financed largely by loans from Rockefeller's Chase Manhattan or Citibank. These same banks made huge windfall profits when, some years later, they organized wealthy Argentines' flight of capital out of the peso into offshore Chase or Citibank "private banking" accounts.

The impact on the general population was anything but positive. With foreign takeovers came massive layoffs of—until then—public workers. Not surprisingly, Argentina's Menem regime, and its economic czar, Domingo Cavallo, were hailed for creating what was labeled in the financial media as the "Argentine Miracle."

Inflation was ended in 1991 by imposing an absolute surrender of monetary control to a Currency Board, a form of central bank whose control was held by the IMF. The Peso, severely devalued from the 1970's level, was rigidly fixed by the Currency Board at 1:1 to the US dollar. No money could be printed nationally to stimulate the economy without an equal increase in dollar reserves in the Currency Board account. The fixed peso opened the floodgates for foreign investors to speculate and reap huge gains on the privatization of the state economy during the 1990's.

When, in April 2001, Cavallo was recalled amid a major economic crisis, to run the national economy once again, he secretly engineered a coup on behalf of the New York banks and his local banking friends. Cavallo simply froze deposits on personal bank accounts of private savers in Argentina to save the assets of his banker friends in New York and elsewhere abroad.

At this point, Argentina defaulted on $132 billion in state debt. Cavallo's first act as Economics Minister in April 2001 was to meet secretly with Rockefeller's JP Morgan-Chase Bank, CSFB's David

Mulford, London's HSBC and a select few other foreign bankers. They swapped $29 billion of old Argentine state bonds for new bonds, a secret deal which made the banks huge profits and which secured their loan exposures to the country. Argentina was the loser as the swap made its total debt burden even larger. A year later Cavallo and the seven foreign banks were subject to judicual investigations that alleged the swaps were illegal and designed to benefit the foreign bankers. According to US financial investors, it actually speeded the default on the state debt. By 2003, total foreign debt had risen to $198 billion, equivalent to three times the level of when Menem took office in 1989.[8]

Rockefeller's Argentina Land Revolution
By the mid-1990's, the Menem government moved to revolutionize Argentina's traditional productive agriculture into monoculture aimed for global export. The script was again written for him in New York and Washington by foreign interests, constituted above all by the associates of David Rockefeller.

Menem argued that the transformation of food production into industrial cultivation of GM soybean was necessary for the country to pay its ballooning foreign debt. It was a lie, but it succeeded in transforming Argentine agriculture into a pawn for North American investors like David Rockefeller, Monsanto and Cargill Inc.

Following almost two decades of economic battering through mounting foreign debts, forced privatization and the dismantling of national protective barriers, the highly-valued Argentine agricultural economy would now be the target of the most radical transformation of them all.

In 1991, several years before field trials were implemented in the United States, Argentina became a secret experimental laboratory for developing genetically engineered crops. The population.was to become the human guinea pigs of the project. Menem's government created a pseudo-scientific Advisory Commission on Biotechnology to oversee the granting of licenses for more than 569 field trials for GM corn, sunflowers, cotton, wheat and especially soybeans.[9] There was no public debate on the initiative of either the Menem government or the Commission on the controversial issue of whether or not GMO crops were safe.

The Commission met in secret, and never made its findings public. It merely acted as a publicity agent for foreign GMO seed multinationals. This was not surprising as the Commission members themselves came from Monsanto, Syngenta, Dow AgroSciences and other GMO giants. In 1996, Monsanto Corporation of St. Louis Missouri was the world's largest producer of genetically-manipulated patented soybean seeds: Roundup Ready soybeans.

In 1995, Monsanto introduced Roundup Ready (RR) soybeans that had a copy of a gene from the bacterium, Agrobacterium sp. strain CP4, inserted, by means of a gene gun, into its genome. That allowed the transgenic or GMO plant to survive being sprayed by the non-selective herbicide, glyphosate. Glyphosate, the active ingredient in Roundup, killed conventional soybeans. Any conventional soybean crops adjacent to Monsanto Roundup Ready crops would inevitably be affected due to wind-borne contamination.[10] Conveniently, that greatly aided the spread of Monsanto crops once introduced.

The genetic modification in Monsanto Roundup Ready soybeans involved incorporating a bacterial version of the enzyme into the soybean plant that gave the GMO soybean protection from Monsanto's herbicide Roundup. Roundup was the same herbicide used by the US Government to eradicate drug crops in Colombia.

Thereby protected, both the soybeans and any weeds could be sprayed with Roundup, killing the weeds and leaving the soybeans. Typically, rather than less herbicide chemicals, GMO soybeans required significantly more chemicals per hectare to control weed growth.[11]

Since the 1970's, soybeans had been promoted by large agribusiness seed companies to become a major source of animal feed worldwide. Monsanto was granted an exclusive license in 1996 by President Menem to distribute its GMO soybean seeds throughout Argentina.

Simultaneous to this wholesale introduction of Monsanto GMO soybean seeds and, necessarily, the required Monsanto Roundup herbicide to Argentine agriculture, now ultra-cheap (in dollar terms), Argentine farmland was bought up by large foreign companies such as Cargill—the world's largest grain and commodity trading company—by international investment funds such as George Soros's Quantum Fund, by foreign insurance companies, and corporate interests such as Seaboard Corporation. This was a hugely profitable operation for foreign investors, for which GMO Monsanto seeds were ultimately the basis for a giant new soy agribusiness industrial farming. Argentina's land was to be converted into a vast industrial seed production unit. For the foreign investors, the beauty of the scheme was that compared with traditional agriculture, GMO soybean needed little human labor.

In effect, as a consequence of the economic crisis, millions of acres of prime farmland were put up for auction by the banks. Typically, the only buyers with dollars to invest were foreign corporations or private persons. Small peasant farmers were offered pennies for their lands. Sometimes, when they refused to sell, they were forced off their properties by terrorist militia or by the state police. Tens of thousands more farmers had to give up their lands when they were driven to bankruptcy by market flooding of cheap food imports brought in under the free market reforms imposed by the IMF.

Additionally, fields planted with the GMO "Roundup Ready" soybean seeds and their special Roundup herbicide required no ordinary turning over of the soil through plowing. In order to maximize profitability, the sponsors of the GMO soybean revolution created huge Kansas-style expanses of land where large mechanized equipment could operate around the clock, often remote-controlled by GPS satellite navigation, without even a farmer needed for driving the tractor.

Monsanto's GMO soybean was sold to Argentine farmers as an ecological plus, utilizing "no-till" farming. In reality they were anything but environmentally friendly.

The GMO soybean and Roundup herbicide were planted with a technique called "direct drilling," pioneered in the USA and with the purpose of saving time and money.[12]

Only affordable to larger wealthy farmers, "direct drilling" required a mammoth special machine which automatically inserts the GM soybean seed into a hole drilled several centimeters deep, and then presses dirt down on top of it. With this direct drilling machine, thousands of acres could be planted by one man. By contrast, a traditional three hectare peach or lemon grove required 70 to 80 farm laborers to cultivate. Previous crop residues were simply left in the field to rot, producing a wide variety of pests and weeds alongside the Monsanto GMO soybean sprouts. That in turn led to greater markets for Monsanto to sell its special patented glyphosate or Roundup herbicide, along with the required Roundup Ready patented soybean seeds. After several years of such planting, the weeds began to show a special tolerance to glyphosate, requiring ever stronger doses of that or other herbicides.[13]

With the decision to license Monsanto genetically engineered Roundup Ready soybeans in 1996, Argentina was to undergo a revolution which its proponents hailed as a "second green revolution." In reality it was the devolution of a once-productive national family farm-based agriculture system into a neo-feudal state system dominated by a handful of powerful, wealthy Latifundista landowners.

The Menem government insured that the door was opened wide to the introduction of GMO soybean seeds. Argentine farmers were in dire economic straits following years of hyperinflation. Monsanto jumped in and extended "credit" to loan-starved farmers to buy Monsanto GMO seeds and Monsanto Roundup herbicide, the only herbicide effective on its Roundup Ready soybean. Monsanto also made the initial transition to GMO soybean more alluring to farmers by offering to provide them with the necessary "direct drilling" machines and training.

"Soybeans for Me, Argentina…"

The results of the GMO soybean revolution in Argentina were impressive in one respect. The nation's agriculture economy was completely transformed in less than a decade.

In the 1970's, before the debt crisis, soybean was not even a factor in the national agriculture economy, with only 9,500 hectares of soybean plantations. In those years, a typical family farm produced a variety of vegetable crops, grains, raised chickens and perhaps a few cows for milk, cheese and meat.

By 2000, after four years of adopting Monsanto soybeans and mass production techniques, over 10 million GMO soy hectares had been planted. By 2004, the area had expanded to more than 14 million hectares. Large agribusiness combines had managed to clear forests, as well as traditional lands occupied by the indigenous people to create more land for soy cultivation.

Argentine agricultural diversity, with its fields of corn, wheat, and cattle, was rapidly being turned into monoculture, just as Egyptian farming was taken over and ruined by cotton in the 1880's.

For more than a century, Argentine farm land, especially the legendary pampas, had been filled with wide fields of corn and wheat amid green pastures grazed by herds of cattle. Farmers rotated between crops and cattle to preserve soil quality. With the introduction of soybean monoculture, the soil, leeched of its vital nutrients, required even more chemical fertilizers—not less, as Monsanto had promised. The large beef and dairy herds which had roamed freely for decades on the grasslands of Argentina were now forced into cramped US-style mass cattle feedlots to make way for the more lucrative soybean. Fields of traditional cereals, lentils, peas and green beans had already almost vanished.

A leading Argentine agro-ecologist, Walter Pengue, a specialist in the impact of GMO soybeans, predicted that, "If we continue in this path, perhaps within 50 years the land will not produce anything at all."[14]

By 2004, 48% of all agricultural land in the country was dedicated to soybean crops, and between 90% and 97% of these were Monsanto GMO Roundup Ready soybeans. Argentina had become the world's largest uncontrolled experimental laboratory for GMO.[15]

Between 1988 and 2003, Argentine dairy farms had been reduced by half. For the first time, milk had to be imported from Uruguay at costs far higher than domestic prices. As mechanized soybean monoculture forced hundreds of thousands of workers off the land, poverty and malnutrition soared.

In the more tranquil era of the 1970's, before the New York banks stepped in, Argentina enjoyed one of the highest living standards in Latin America. The percentage of its population officially below the poverty line was 5% in 1970. By 1998, that figure had escalated to 30% of the total population. And by 2002, to 51%. By 2003, malnutrition rose to levels estimated at between 11% and 17% of the total population of 37 million.[16]

Amid the drastic national economic crisis arising from the state's defaulting on its debt, Argentines found they were no longer able to rely on small plots of land for their survival. The land had been overrun by mass GMO soybean acreages and blocked to even ordinary survival crops.

Under the support of foreign investors and agribusiness giants like Monsanto and Cargill, large Argentine landowners moved systematically to seize land from helpless peasants, most often with backing from the state. By law, peasants had rights over lands of which they had the uncontested use for 20 years or more. That traditional right was trampled by the powerful new interests behind agribusiness. In the vast region of Santiago del Estero in the north, large feudal landowners began an operation of mass deforestation to make way for wholesale GMO soybean crops.

Peasant communities were suddenly told that their land belonged to someone else. Typically, if they refused to leave willingly, armed groups would steal their cattle, burn their crops and threaten them with more violence. The lure of huge profits from GMO soybean exports was the driving force behind the violent upheaval surrounding traditional farming across the country.

As farming families were made destitute and pushed off their lands, they fled to new shanty towns on the edges of the larger cities, turning to social disorder, crime and suicide, while disease became rampant amid the impossible overcrowding. Within several years, more than 200,000 peasants and small farmers were driven off their lands to make way for the large agribusiness soybean planters.[17]

Monsanto Conquers with Deception

Taking the example of the old 16th Century Spanish Conquistadores, Monsanto's warriors conquered the land with a campaign of lies and deception. Because Argentina's national Seed Law did not protect Monsanto's patent on its glyphosate-resistant genetically modified soybean seed, the company could not legally demand a patent royalty when Argentine farmers reused their soybean seeds in the next harvest season. Indeed, not only was it traditional, but also legal, for Argentine farmers to re-plant seeds for their own use.

Collection of such a royalty or "technology license fee" was at the heart of the Monsanto marketing scheme. Farmers in the USA and elsewhere had to sign a binding contract with Monsanto agreeing to not re-use saved seeds and to pay new royalties to Monsanto each year—a system which can be seen as a new form of serfdom.

To get around the refusal by the nationalist Argentine Congress to pass a new law granting Monsanto the right to impose royalty payments against severe court-imposed fines, Monsanto adopted another ploy. Farmers were sold the initial seeds needed to expand the soybean revolution in Argentina. In this early stage, Monsato deliberately waived its "technology license fee," favoring the widest possible proliferation of its GM seeds across the land, and in particular, of the patented glyphosate Roundup herbicide that went along with it. The insidious marketing strategy behind selling glyphosate-resistant seeds was that farmers were forced to purchase the specially matched Monsanto herbicides.

GMO soybean planted land increased 14-fold, while the smuggling of Monsanto Roundup soybean seeds spread across the Pampas and into Brazil, Paraguay, Bolivia and Uruguay. Monsanto did nothing to stop what it saw as the illegal spread of its seeds.[18] Monsanto partner Cargill was itself accused of illegally smuggling GMO soybean seeds secretly mixed with non-GMO seeds, into Brazil from Argentina. Amusingly, in Brazil, the smuggled Argentine GMO soybean seeds were called "Maradona" seeds in reference to the famous Argentine football player later treated for cocaine addiction.

Finally, in 1999, three years after its introduction of GMO soybeans, Monsanto formally demanded farmers to pay up the "extended royalties" on the seeds, despite the fact that Argentine law made it illegal to do so. The Menem government made no protest against Monsanto's brazen orders, while farmers ignored it altogether. But the stage was being set for the next legal act. Monsanto claimed the royalties were necessary for it to recover its investments on the "research and development" of the GMO seeds. It began a careful public relations campaign designed to paint itself as the victim of farmers' abuse and "theft".

In early 2004, Monsanto escalated its pressure on the Argentine government. Monsanto announced that if Argentina refused to recognize the "technology license fee," it would enforce its collection at points of import such as the USA or the EU, where Monsanto patents were recognized, a measure which would spell a devastating blow to the market for Argentine agribusiness exports. Moreover, after Monsanto's well-publicized threat to stop selling all GMO soybeans in Argentina, and the claim that more than 85% were illegally replanted by farmers in what was branded a "black market," the Agriculture Secretary, Miguel Campos, announced that the government and Monsanto had come to an agreement.

A Technology Compensation Fund was to be created and managed by the Ministry of Agriculture. Farmers would have to pay a royalty or tax fee of up to almost one percent on the sale of GMO soybeans to grain elevators or exporters such as Cargill. The tax was to be collected at the processing site, leaving farmers with no choice but to pay up if they were to process their harvest. The tax would then be paid back to Monsanto and other GMO seed suppliers by the government.[19]

Despite fierce farmer protest, the Technology Compensation Fund was implemented at the end of 2004.

By early 2005, the Brazilian government of President Luiz Inacio Lula da Silva had also thrown in the towel, and passed a law making planting of GMO seeds in Brazil legal for the first time, claiming that the use of GMO seeds had spread so widely as to be uncontrollable anyway. The barriers to GMO proliferation across Latin

America were melting. By 2006, together with the United States, where GMO Monsanto soybeans dominated, Argentina and Brazil accounted for more than 81% of world soybean production, thereby ensuring that practically every animal in the world fed soymeal was eating genetically engineered soybeans. Similarly, this would imply that every McDonald's hamburger mixed with soymeal would be genetically engineered, and most processed foods, whether they realized or not.[20]

Let Them Eat Soybeans!

As the GMO soybean revolution destroyed traditional agricultural production, Argentines faced a dramatic change in their available diet. Furthermore, widespread soybean-based monoculture left the population desperately vulnerable to the national economic depression which hit Argentina in 2002. Previously in tough times, farmers and even ordinary city dwellers could grow their own crops to survive. But that was no longer possible under the transformation of Argentina's agriculture into industrial agribusiness.

As a result, hunger spread across the land, just as the economic crisis worsened. Fearing food riots, the national government, aided by Monsanto and the giant international soybean users such as Cargill, Nestlé, and Kraft Foods, responded by giving out free food to the hungry. Meals made from soybeans were thus distributed with the secondary motive of fostering wider domestic consumption of the crop.

A national campaign was put in motion urging Argentines to replace a healthy diet of fresh vegetables, meat, milk, eggs and other products with ... soybeans. DuPont AgriSciences created a new organization with the healthy-sounding name, "Protein for Life," in order to propagate soybean consumption by humans, even though the soybeans were meant to be grown as animal feed. As part of the campaign, DuPont gave out food fortified with soybeans to thousands of Buenos Aires poor. It was the first time ever in any country that a population had directly consumed soybeans in such large quantities. The Argentines had now become guinea pigs in more ways than one.[21]

Government and private propaganda touted the great health benefits of a soybean diet, as a replacement for dairy or meat protein. But the campaign was based on lies. It conveniently omitted the fact that a diet based on soybean is unfit for long-term human consumption, and that studies have established that babies fed soymilk have dramatically higher levels of allergies than those fed breast milk or cow milk. They did not tell Argentines that raw and processed soybeans contain a series of toxic substances which, when soy is consumed as a staple element of one's diet, damage health and have been related to cancer. They refused to say that soybeans contain an inhibitor, Trypsin, which Swedish studies have linked to stomach cancer.[22]

In the countryside, the impact of mass soybean monoculture was horrendous. Traditional farming communities close to the huge soybean plantations were seriously affected by the aerial spraying of Monsanto Roundup herbicides. In Loma Senes, peasants growing mixed vegetables for their own consumption found all their crops destroyed by spraying, as Roundup kills all plants other than specially gene-modified "herbicide-resistant" Monsanto beans.

A study conducted in 2003 showed that the spraying had not only destroyed the nearby peasants' crops: their chickens had died and other animals, especially horses, were adversely affected. Humans contracted violent nausea, diarrhea, vomiting and skin lesions from the herbicide. There were reports of animals born near GMO soybean fields with severe organ deformities, of deformed bananas and sweet potatoes, of lakes suddenly filled with dead fish. Rural families reported that their children developed grotesque blotches on their bodies after the spraying of nearby soybean fields.

Added damage occurred to valuable forest land, which was bulldozed to make way for mass-cultivation of soybean, especially in the Chaco region near Paraguay and the Yungas region. The loss of forests created an explosion of cases of medical problems among indigenous inhabitants, including leishmaniasis, a parasite transmitted by sand flies, which is expensive to treat and leaves severe scars and other deformities. In Entre Rios, more than 1.2 million acres of forest were removed by 2003, at which point the government finally issued an order forbidding further deforestation.

To convince wary Argentine farmers to use Monsanto Roundup Ready soybean seeds in 1996, the company had made grand claims of a miracle crop, arguing that its GMO soybean was genetically modified to be resistant to Monsanto's Roundup herbicide.

The company assured farmers that they would therefore require dramatically less herbicide and chemical treatment for their soybean crops than with regular soybean. As Roundup kills virtually everything that grows aside from Monsanto GMO soybeans, only one, rather than several, herbicides would be necessary—or so went Monsanto's PR campaign. Grand promises were also made about higher yields and lower costs, feeding the desperate farmers with dreams of a better economic situation. Not surprisingly, the response was hugely positive.

On average, the Roundup soybean crops gave between 5% to 15% lower yields than traditional soybean crops. Also, far from needing less herbicide, farmers found vicious new weeds which needed up to three times as much spraying as before. The United States Department of Agriculture (USDA) statistics from 1997 showed that expanded plantings of Roundup Ready soybeans resulted in a 72% increase in the use of glyphosate.[23]

According to the Pesticides Action Network, scientists estimated that plants genetically engineered to be herbicide resistant will actually triple the amount of herbicides used. Farmers, knowing that their crop can tolerate or resist being killed off by the herbicides, will tend to use them more liberally. Monsanto never conducted rigorous independently verifiable tests of the negative health effects of feeding cattle, let alone humans, with the raw Monsanto soybeans saturated with Roundup herbicides. The increased use of chemicals led to larger costs than with non-GMO seeds.[24]

But by the time the farmers realized this, it was too late. By 2004, GMO soybean had spread across the entire country, and the seeds all depended on Monsanto Roundup pesticide. A more perfect scheme of human bondage would be hard to imagine.

Yet Argentina was not the only target land for the project of gene-manipulated agriculture crops. The Argentine case was but the first stage in a global plan that was decades in the making and absolutely shocking and awesome in its scope.

Notes

1. Pew Initiative on Food and Biotechnology, *Genetically Modified Food Crops in the United States*, http://www.pewagbiotech.org, August 2004.

2. F. William Engdahl, *A Century of War: Anglo-American Oil Politics and the New World Order*, Pluto Books Ltd., London, 2004, Chapters 10-11. John Perkins, *The Confessions of an Economic Hit Man*, Berrett-Koehler Publishers, San Francisco, 2004.

3. U.S. Embassy, *Document #1976 Buenos06130*, 20 September 1976, part of declassified US State Department documents. Cynthia J. Arnson (editor), *Argentina-United States Bilateral Relations*, Woodrow Wilson Center for Scholars, Washington D.C., 2003, pp. 39-40. Kissinger's conversation with Guzzetti in Santiago was first reported by Martin Edwin Andersen, "Kissinger and the Dirty War", *The Nation*, 31 October 1987. Andersen's article was based on a memo by Assistant Secretary for Human Rights Patricia Derian, who was told the story by Hill during a visit to Argentina in March 1977. Hill demarche on human rights: Buenos Aires 3462, May 25, 1976, "Request for Instructions", State 129048, 25 May 1976, "Proposed Demarche on Human Rights."

4. Francisco J. Ruge-Murcia, *Heterodox Inflation Stabilization in Argentina, Brazil and Israel*, Centre de recherche et développement en économique (C.R.D.E.) and Département de sciences économiques, Université de Montréal, May 1997.

5. Asad Ismi, "Cry for Argentina", *Briarpatch*, September 2000.

6. David Rockefeller, "Lo que pienso de Martínez de Hoz", *Revista Gente*, 6 April 1978.

7. Government of Argentina Ministry of Education, *La Dictadura Militar en Argentina:24 de marzo de 1976—10 de diciembre de 1983*, http://www.me.gov.ar/efeme/24demarzo/dictadura.html, 2001. Cavallo was indicted in 2006 by the Government of Argentina for knowingly conspiring with US banker Mulford in a 2001 debt swap that was declared "fraud" and cost Argentina tens of billions more in debt servicing to Mulford and other creditor banks. That swap led to the Argentine default later in 2001. Details in MercoPress, *Former Argentine Leader Indicted for 2001 Bond Swap*, http://www.mercopress.com. Details of the debt fraud are also well described in Jules Evans, *Bankers Accused of Dirty Tricks in Argentina*, http://www.euromoney.com, 28 January 2002.

8. Jules Evans, *Bankers Accused of Dirty Tricks in Argentina*, 28 January 2002, http://www.euromoney.com/public/markets/banking/news/30jan02-1.html.

9. Canadian Market Research Centre Market Support Division (TCM) Department of Foreign Affairs and International Trade, *Market Brief: The Biotechnology Market in Argentina: Government Support for Biotechnology*, May 2003, http://www.ats.agr.gc.ca/latin/3720_e.htm.

10. American Chemical Society, "Growing Evidence of Widespread GMO Contamination", *Environmental Science & Technology: Environmental News*, 1 December 1999, Vol. 33, No. 23, pp. 484 A-485 A.

11. Judy Carman, *The Problem with the Safety of Roundup Ready Soybeans*, Flinders University, Southern Australia, http://www.biotech-info.net, August 1999.

12. UK Soil Management Initiative, *Frequently Asked Questions: Advantages and Disadvantages of Minimum Tillage*, http://www.smi.org.uk.

13. *Ibid.*

14. Sue Branford, "Argentina's Bitter Harvest", *New Scientist*, 17 April 2004, pp. 40-43. See also Organic Consumers Association, *New Study Links Monsanto's Roundup to Cancer*, 22 June 1999, Little Marais, MN.

15. Lillian Joensen and Stella Semino, "Argentina's Torrid Love Affair with the Soybean", *Seedling*, October 2004, p. 3. This is an excellent summary of the interplay between the foreign debt crisis, the IMF policies of privatization, and the transformation of Argentine agriculture by GMO seeds. The authors are with the Rural Reflection Group, in Argentina.

16. *Ibid.*, p. 4.

17. *Ibid.*, p. 3.

18. Lillian Joensen, *op. cit.*, p. 3.

19. GRAIN, *Monsanto's Royalty Grab in Argentina*, http://www.grain.org, October 2004.

20. Sue Branford, "Why Argentina Can't Feed Itself," *The Ecologist*, October 2002. H. Paul, R. Steinbrecher, et al., *Argentina and GM Soybean: The Cost of Complying with US Pressure*, EcoNexusBriefing, 2003, http://www.econexus.info. David Jones, "Argentina and GM Soy—Success at What Cost?" *Saturday Star*, South Africa, 19 June 2004.

21. Lillian Joensen, *op.cit.*, p. 5.

22. Lennart Hardell, Miikael Eriksson, "A Case-Control Study of Non-Hodgkin Lymphoma and Exposure to Pesticides", *Cancer*, 15 March 1999. A joint USA-New Zealand independent research organization, SoyOnlineService, states that contrary to widely promoted claims of health and dietary benefits, "[s]oy foods contain trypsin inhibitors that inhibit protein digestion and affect pancreatic function. In test animals, diets high in trypsin inhibitors led to stunted growth and pancreatic disorders. Soy foods increase the body's requirement for vitamin D, needed for strong bones and normal growth. Phytic acid in soy foods results in reduced bioavailabilty of iron and zinc which are required for the health and development of the brain and nervous system. Soy also lacks cholesterol, likewise essential for the development of the brain and nervous system. Megadoses of

phytoestrogens in soy formula have been implicated in the current trend toward increasingly premature sexual development in girls and delayed or retarded sexual development in boys...Soy isoflavones are phyto-endocrine disrupters. At dietary levels, they can prevent ovulation and stimulate the growth of cancer cells. Eating as little as 30 grams (about 4 tablespoons) of soy per day can result in hypothyroidism with symptoms of lethargy, constipation, weight gain and fatigue." in *Myths & Truths About Soy Foods* printed in SoyOnlineService.co.nz.

23. Cited in Royal Society of New Zealand, *Genetic Engineering—an Overview, 4. Environmental Aspects of Genetic Engineering,* in http://www.rsnz.org/topics/biol/gmover/4.php.

24. Genetic Concern, *New Study Links Monsanto's Roundup to Cancer,* June 1999, in http://www.biotech-info.net/glyphosate_cancer.html.

CHAPTER 10

Iraq Gets American Seeds
of Democracy

> "The reason we are in Iraq is to plant the seeds of
> democracy so they flourish there and spread to the
> entire region of authoritarianism."
>
> *George W. Bush*

US-style Economic Shock Therapy

When George W. Bush spoke of planting the "seeds of democracy" few realized that he had Monsanto genetically-engineered seeds in mind.

Following the US occupation of Iraq in March 2003, the economic and political realities of that country changed radically. Not only was Iraq occupied by some 130,000 US troops and a small army of private mercenary soldiers of fortune closely tied to the Pentagon, it was also under the comprehensive economic control of its occupier, the United States.

Control over the Iraqi economy was run out of the Pentagon. In May 2003, Paul Bremer III was put in charge as Administrator of the newly created Coalition Provisional Authority, or CPA, a thinly-veiled occupation authority. Bremer, a former US State Department terrorism official, had gone on to become Managing Director of the powerful consulting firm of former Secretary of State Henry Kissinger, Kissinger Associates.

In many respects, US-occupied Iraq was a far better opportunity than Argentina. The US occupation was instrumental in bringing the agricultural system of an entire country under the domain of GMO agribusiness. The US occupation administration simply made Iraqi farmers an offer they could not refuse: "Take our GM seeds or die."

Bremer held de facto life-and-death control over every area of civilian activity in occupied Iraq. Notably, he did not report to the State Department, which is typically the department responsible for reconstruction, but directly to the office of former Defense Secretary Donald Rumsfeld, in the Pentagon.

As head of the CPA, Bremer moved swiftly to draft a series of laws to govern Iraq, which at the time had neither a constitution nor a legally-constituted government. The new laws of the US occupation authority numbered 100 in all, and were put into effect in April 2004.[1] As a whole, the hundred new US-mandated laws—or orders, as they were called—would insure that the economy of Iraq would be remade along the lines of a US-mandated free-market economic model; much as the International Monetary Fund and Washington had imposed on the economies of Russia and the former Soviet Union after 1990.

The mandate given to Bremer by Rumsfeld's Pentagon planners, was to impose a "shock therapy" that would turn the entire state-centered economy of Iraq into a radical free-market private region. He executed more drastic economic changes in one month than were forced on the debtor countries of Latin America in three decades.

Bremer's first act was to fire 500,000 state workers, most of them soldiers, but also doctors, nurses, teachers, publishers, and printers. Next, he opened the country's borders to unrestricted imports: no tariffs, no duties, no inspections, no taxes. Two weeks after Bremer came to Baghdad in May 2003, he declared Iraq to be "open for business." He did not say whose business, but that was becoming increasingly clear.

Before the invasion, Iraq's non-oil economy had been dominated by some 200 state-owned companies, which produced everything from cement to paper to washing machines. In June 2003, Bremer announced that these state firms would be privatized immediately.

"Getting inefficient state enterprises into private hands," he said, "is essential for Iraq's economic recovery."[2] The Iraqi privatization plan would be the largest state liquidation sale since the collapse of the Soviet Union.

CPA Order 37 lowered Iraq's corporate tax rate from roughly 40 percent to a flat 15 percent. Without tax revenues, the state would be unable to play a large role in anything. Order 39 allowed foreign companies to own 100 percent of Iraqi assets outside the natural-resource sector. This ensured unrestricted foreign business activities in the country. Investors could also take 100 percent of the profits they made in Iraq out of the country. They would not be required to reinvest and they would not be taxed. The beneficiaries of these laws were clearly not the people or the economy of Iraq.

Under Order 39, the foreign companies could sign leases and contracts that would last for forty years. Order 40 welcomed foreign banks to Iraq under the same favorable terms. Appropriate to such a foreign takeover of the economy, the only laws remaining from Saddam Hussein's era were those restricting trade unions and collective bargaining.

Overnight, Iraq went from being the most isolated country in the world, to being the freest and most wide-open market. With its economy and banking system devastated by war and more than a decade of US-led economic embargo, Iraqis were in no position to buy their privatized state companies. Foreign multinationals were the only possible actors who might benefit from Bremer's grand economic recovery scheme.

The new laws were imposed on a conquered and devastated land, with no possibility of objection aside from military sabotage and guerrilla warfare against the occupiers. Enacted by the United States Government occupying agency, the CPA, to make Iraq attractive to foreign investors, the set of 100 new orders gave total rights and control over the economy of Iraq to multinational corporations.

Moreover, these laws were designed to pave the way for the most radical transformation of a nation's food production system ever attempted. Under Bremer, Iraq was to become a model for Genetically Modified or GMO agribusiness.

Bremer's Order 81

The CPA explicitly defined the legal magnitude of the 100 Orders. An Order was defined as "binding instructions or directives to the Iraqi people that create penal consequences or have a direct bearing on the way Iraqis are regulated, including changes to Iraqi law." In other words, Iraqis were told "do it or die." Whenever prior Iraqi law might interfere with Bremer's new 100 Orders, Iraqi law was made null and void. The law of occupation was supreme.[3]

Buried deep among the new Bremer decrees, which dealt with everything from media to privatization of state industries, was Order 81 on "Patent, Industrial Design, Undisclosed Information, Integrated Circuits and Plant Variety Law." Order 81 stated:

> 11. Article 12 is amended to read as follows: "A patent shall grant its owner the following rights:
>   1. Where the subject of the patent is a product, the right to prevent any person who has not obtained the owner's authorization from making, exploiting, using, offering for sale, selling or importing that product."
> 12. Article 13.1 is amended to read as follows: "The term of duration of the patent shall not end before the expiration of a period of twenty years for registration under the provisions of this Law as from the date of the filing of the application for registration under the provisions of this Law."

A further provision of Order 81 stated, "Farmers shall be prohibited from re-using seeds of protected varieties or any variety mentioned in items 1 and 2 of paragraph (C) of Article 14 of this Chapter." Furthermore,

> CPA Order No. 81 amends Iraq's patent and industrial design law to protect new ideas in any field of technology that relates to a product or manufacturing processes. The amendments permit companies in Iraq, or in countries that are members of a relevant treaty to which Iraq is a party, to register patents in Iraq. The amendments grant the patent owner the right to prevent any person who has not obtained the owner's authorization from exploiting the patented product or process for twenty years from the date of the patent's

registration in Iraq. The amendments also allow individuals and companies to register industrial designs.[4]

In plain English, Order 81 gave holders of patents on plant varieties (which all happened to be large foreign multinationals) absolute rights over use of their seeds in Iraqi agriculture for 20 years. While that might appear to be a fair and sensible business provision to compensate a foreign company for its intellectual property, in reality it was an incursion on the sovereignty of Iraq. Like many countries, Iraq never recognized the principle of commercial patents on life forms such as plants. The patents had been granted to companies like Monsanto or DuPont by US or other foreign patent authorities.

What Order 81 did, in fact, was amend Iraq's patent law to recognize foreign patents, regardless of the legality of such patents under Iraq law. On the surface, it appeared to leave Iraqi farmers the option to refuse to buy Monsanto or other patented seeds, and to plant their traditional native seeds. In reality, as the drafters of Order 81 were also well aware, it had a quite opposite effect.

The protected plant varieties were Genetically Modified or Gene Manipulated plants, and Iraqi farmers who chose to plant such seeds were required to sign an agreement with the seed company holding the patent, stipulating that they would pay a "technology fee" and an annual license fee for planting the patented seeds.

Any Iraqi farmer seeking to take a portion of those patented seeds to replant in following harvest years would be subject to heavy fines from the seed supplier. In the United States, until a Court ruling struck it down, Monsanto demanded a punitive damage equal to 120 times the cost of a bag of its GMO seeds. This was the occasion for Iraqi farmers to become vassals not of Saddam Hussein, but of the multinational GM seed giants.

At the heart of Order 81 was the Plant Variety Protection (PVP) provision. Under the PVP, seed saving and reuse would become illegal. Farmers using patented seeds or even "similar" seeds, would be subject to severe fines or even prison. However, the plant varieties being protected were not those which resulted from 10,000 years of

Iraqi farm cross-breeding and development. Rather, the protection was given to back up the rights of giant multinational companies to introduce their own seeds and herbicides into the Iraqi market with full protection of the government, both of the US and of Iraq.

Iraqi Seed Treasure Destroyed

Iraq is historically part of Mesopotamia, the cradle of civilization, where for millennia the fertile valley between the Tigris and Euphrates rivers created ideal conditions for crop cultivation. Iraqi farmers have been in existence since approximately 8,000 B.C., and developed the rich seeds of almost every variety of wheat used in the world today. They did this through a system of saving a share of seeds and replanting them, developing new naturally resistant hybrid varieties through the new plantings.

For years, the Iraqis held samples of such precious natural seed varieties in a national seed bank, located in Abu Ghraib, the city better known internationally as the site of a US military torture prison. Following the US occupation of Iraq and its various bombing campaigns, the historic and invaluable seed bank in Abu Ghraib vanished, a further casualty of the Iraq war.

However, Iraq's previous Agriculture Ministry had taken the precaution to create a back-up seed storage bank in neighboring Syria, where the most important wheat seeds are still stored in an organization known as the International Center for Agricultural Research in Dry Areas (ICARDA), based in Aleppo, Syria. With the loss of Abu Ghraib's seed bank, ICARDA, a part of the international Consultative Group on International Agritultural Research (CGIAR) network of seed banks, could have provided the Iraqis with seeds from its store had the CPA wanted to request such help.[5] It did not. Bremer's advisers had different plans for Iraq's food future.

Iraqi agriculture was to be "modernized," industrialized, and reoriented away from traditional family multi-crop farming into US-style agribusiness enterprises producing for the "world market." Serving the food security needs of hungry Iraqis would be incidental to the plan.

Under Bremer's Order 81, if a large international corporation developed a seed variety resistant to a particular Iraqi pest, and an Iraqi farmer was growing another variety that did the same, it was illegal for the farmer to save his own seed. Instead, he is obliged to pay a royalty fee for using Monsanto's GMO seed.

In the late 1990's, a US biotech company, SunGene, patented a sunflower variety with very high oleic acid content. It did not merely patent the genetic structure though. It patented the characteristic of high oleic content itself, claiming right to it. SunGene informed other sunflower breeders that should they develop a variety "high in oleic acid," that it would be considered an infringement of the patent.

"The granting of patents covering all genetically engineered varieties of a species ... puts in the hands of a single inventor the possibility to control what we grow on our farms and in our gardens," remarks Dr. Geoffrey Hawtin, Director General of the International Plant Genetic Resources Institute. "At the stroke of a pen, the research of countless farmers and scientists has potentially been negated in a single, legal act of economic highjack."[6] Economic hijack was just what Bremer and Monsanto intended for Iraq under Order 81.

Such total control on farmer seed varieties was possible under the new law on patent rights in Iraq. The CPA's Order 81, behind the cover of complicated legal jargon, effectively turned the food future of Iraq over to global multinational private companies— hardly the liberation most Iraqis had hoped for.

The patent laws on plants decreed by Order 81, unlike other national laws on Intellectual Property Rights, were not negotiated between sovereign governments or with the WTO. They were imposed by Washington on Iraq without debate. According to informed Washington reports, the specific details of Order 81 on plants were written for the US Government by Monsanto Corporation, the world's leading purveyor of GMO seeds and crops.

No Seeds to Plant

On paper, it appeared that only those seeds which Iraqi farmers chose to buy from international seed companies would fall under the new US-imposed Iraqi law on patents. In reality, Iraq was being turned into a huge laboratory for the development of food products under the control of GMO seed and chemical giants such as Monsanto, DuPont and Dow.

In the aftermath and devastation of the Iraq war, most Iraqi farmers were forced to turn to their Agriculture Ministry for new seeds if they were to plant ever again. Here was the opening for Bremer's takeover of the Iraqi food supply.

For over a decade, Iraqi farmers had endured the US-UK-led embargo on much needed agricultural equipment. Also, Iraq had suffered from three years of severe drought prior to the war, a climatic misfortune which caused Iraqi wheat crops to decline severely. Years of war and economic embargo had thus already devastated Iraqi agriculture and by 2003 grain production had fallen to less than half the level of 1990 before the first Iraq-US war. Up to 2003, much of the Iraqi population had depended on UN oil-for-food rations to survive.

In the name of "modernizing" Iraqi food production, the United States Agency for International Development (USAID) and the US Agricultural Reconstruction and Development Program for Iraq (ARDI) stepped in to transform traditional Iraqi agriculture. The key Washington-appointed agriculture czar for Iraq at that time was Daniel Amstutz, former US Department of Agriculture official and former Vice President of the giant grain conglomerate Cargill Corporation. Amstutz was one of the key persons who had crafted the US demands on Agriculture during the GATT Uruguay Round which led to the creation of the World Trade Organization (WTO) in 1995.

The alleged aim of Order 81 was "to ensure good quality seeds in Iraq and to facilitate Iraq's accession to the World Trade Organization." "Good quality" was of course to be defined by the occupation authority. WTO accession meant Iraq had to open its markets and laws to rules dictated by the powerful industrial and financial interests dominating WTO policy.

As soon as Order 81 was issued, USAID began delivering, through the Agriculture Ministry, thousands of tons of subsidized,

US-origin "high-quality, certified wheat seeds" to desperate Iraqi farmers that were initially nearly cost-free. According to a report by GRAIN, an NGO critical of GMO seeds and plant patents, USAID refused to allow independent scientists to determine whether or not the seed was GMO. Naturally, should it prove to be GMO wheat seed, within one or two seasons, Iraqi farmers would find themselves obligated to pay royalty fees to foreign seed companies in order to survive. The GRAIN report stated the intent of Order 81:

> The CPA has made it illegal for Iraqi farmers to re-use seeds harvested from new varieties registered under the law. Iraqis may continue to use and save from their traditional seed stocks or what's left of them after the years of war and drought, but that is the not the agenda for reconstruction embedded in the ruling. The purpose of the law is to facilitate the establishment of a new seed market in Iraq, where transnational corporations can sell their seeds—genetically modified or not, which farmers would have to purchase afresh every single cropping season.[7]

While historically Iraq prohibited private ownership of biological resources, the new US-imposed patent law introduced a system of monopoly rights over seeds—rights which no Iraqi farmer had the resources to develop.

In effect, Bremer inserted into Iraq's previous patent law a new chapter on Plant Variety Protection (PVP) that was said to provide for the "protection of new varieties of plants." PVP, an Intellectual Property Right (IPR), was in fact a patent for plant varieties which gave exclusive rights on planting materials to a plant breeder who claimed to have discovered or developed a new variety.

The protection in the PVP had nothing to do with conservation, but referred to "safeguarding of the commercial interests of private breeders." Under the US decree, "plant variety protection" really spelt plant variety destruction.

"Let Them Eat... Pasta?"

Under the program, the State Department, working with the US Department of Agriculture (USDA), had set up 56 "wheat extension demonstration sites" in northern Iraq with the purpose of "introducing and demonstrating the value of improved wheat seeds." The project was run for the US Government by the International Agriculture Office of Texas A&M University, which used its 800 acres of demonstration plots all across Iraq to teach farmers how to grow "high-yield seed varieties" of crops that included barley, chick peas, lentils and wheat.[8]

The $107 million USAID agriculture reconstruction project's goal was to double the production of 30,000 Iraqi farms within the first year. The idea was to convince skeptical Iraqi farmers that only with such new "wonder seeds" could they get large harvest yields. As had been the case ten years earlier with American farmers, desperation and a promise of huge gains would be used to trap Iraqi farmers into dependence on foreign seed multinationals.

Coincidentally, Texas A&M's Agriculture Program also described itself as "a recognized world leader in using biotechnology," or GMO technology. With their new seeds would come new chemicals—pesticides, herbicides, fungicides, all sold to the Iraqis by corporations such as Monsanto, Cargill and Dow.

The Business Journal of Phoenix, Arizona, reports that "An Arizona agri-research firm is supplying wheat seeds to be used by farmers in Iraq looking to boost their country's home-grown food supplies." That firm was called the World Wide Wheat Company (WWWC), and in partnership with three universities, including Texas A&M, it would "provide 1,000 pounds of wheat seeds to be used by Iraqi farmers north of Baghdad."[9]

According to Seedquest, a central information website for the global seed industry, WWWC was a leader in developing "proprietary varieties" of cereal seeds—i.e., varieties that are patented and owned by a particular company.[10] These were the sorts of protected GMO seeds contained in the Order 81. According to WWWC, any "client," or farmer as they were once known, who wishes to grow one of their seeds "pays a licensing fee for each variety." W3, as it called itself, formally works in cooperation with the Bio5 Institute of biosciences at the University of Arizona, which curiously describes itself as a "state-of-the-art garage for bio-research."[11]

Even more remarkable, according to the Phoenix *Business Journal* article, "six kinds of wheat seeds were developed for the Iraqi endeavor. Three will be used for farmers to grow wheat that is made into pasta; three seed strains will be for bread-making."[12] That meant that 50% of the grains being developed by the US in Iraq after 2004 were intended for export. Indeed, pasta was a food fully foreign to the Iraqi diet, demonstrating that, rather than to produce food for the starving 25 million war-weary Iraqis, Bremer's Order 81 was designed to create industrialized agro-business using GM seeds for production geared toward global export.

Additionally, the $107 million USAID agricultural reconstruction project had the aim to get the Iraqi government out of food production. "The idea is to make this a completely free market," said Doug Pool, agriculture specialist with USAID's Office of Iraq Reconstruction.[13]

The USAID aim—mirroring US and WTO policies—was to help the new government phase out farm subsidies. "The Minister of Agriculture has been quite good in doing that," Pool said. State enterprises, such as the Mesopotamia Seed Co., "need to be spun off and privatized," he declared.[14] He did not mention who would have the cash in war-torn Iraq to buy such a state seed company. Only rich foreign agribusiness giants such as Monsanto could be likely buyers.

To facilitate the introduction of patent-protected GM seeds from foreign seed giants, the Iraq Agriculture Ministry distributed these GM seeds at "subsidized prices." Once farmers started using the GM seeds, under the new Plant Patent Protection rules of Order 81, they would be forced to buy new seeds each year from the company. Under the banner of bringing a "free-market" into the country, Iraqi farmers were becoming enslaved to foreign seed multinationals.

In a December 2004 interview, Iraq's US-educated interim Agriculture Minister, Sawsan Ali Magid al-Sharifi, stated, "We need Iraqi farmers to be competitive, so we decided to subsidize inputs like pesticides, fertilizers, improved seeds and so on. We cut down on the other subsidies, but we have to become competitive."[15]

In other words, money for Iraq's impoverished farmers to buy new seeds was earmarked for buying GMO "improved seeds" from foreign multinationals like Monsanto.

At the same time, US commodity exporters were hungrily eyeing new market opportunities. "Iraq was once a significant commercial market for US farm products, with sales approaching $1 billion in the 1980s," told Bush Administration former agriculture secretary Ann Veneman—who had ties to Monsanto before she came to Washington—to a conference of farm broadcasters in 2003. "It has the potential, once again, to be a significant commercial market."[16]

What Veneman neglected to say was that during the Iran-Iraq war in the late 1980's, the Reagan and Bush Administrations disguised arms and chemical weapons sales to Saddam Hussein's Iraq under the US Department of Agriculture Commodity Credit Corporation export program. The scandal involved billions of US taxpayer dollars and implicated former Secretary of State Henry Kissinger and National Security Adviser, Brent Scowcroft, as well as the Atlanta branch of the Italian Banco Nazionale de Lavoro (BNL).[17]

According to John King, vice chairman of the US Rice Council, Iraq was the top market for US rice in the late '80s, prior to the 1991 Gulf war. "The US rice industry wants to play a major role once again in supplying rice to Iraq," King told the US House Agriculture Committee. "With the current challenges facing the US rice industry ... renewed Iraqi market access could have a tremendous impact in value-added sales."[18]

King added that, "The liberation of Iraq in 2003 by coalition forces has brought freedom to the Iraqi people.[19] The resumption of trade has also provided hope for the US rice industry." He failed to mention that, in 2003, most US rice was genetically manipulated.

In Spring 2004 as Order 81 was promulgated by Bremer's CPA, supporters of the radical young cleric Moqtada al Sadr protested the closing of their newspaper, *al Hawza*, by US military police. The CPA accused *al Hawza* of publishing "false articles" that could "pose the real threat of violence." As an example, the CPA cited an article that claimed Bremer was, "pursuing a policy of starving the Iraqi people to make them preoccupied with procuring their daily bread so they do not have the chance to demand their political and individual freedoms."[20]

That such articles would appear in light of Order 81 was hardly surprising. Neither was it surprising that Bremer's CPA would vigorously try to silence such criticisms of its food policy given the stakes for the entire GMO project.

Iraq, USA and the IMF Dictates

On November 21, 2004, the leading representatives of the Paris Club of creditor governments issued a proclamation on how they would handle the estimated $39 billion Iraqi government debt owed to the industrial countries at large, as part of the estimated $120 billion foreign debts from the Saddam Hussein era. Despite the overthrow of the regime of Saddam Hussein, Washington was initially not about to wipe the slate clean and declare the old debts illegitimate.

The Paris Club governments agreed to new terms on the limited $39 billion state-owed debts only after heavy pressure from US Iraq Special Debt Negotiator, James Baker III. Baker was no novice negotiator. He engineered the election of George W. Bush in 2001 through an appeal to the Supreme Court, and he is one of the closest advisers of the Bush family.

In the ensuing horse trading with its OECD allies, the US Government was quite happy to press for a major write-off of old Iraqi debt to Paris Club creditors, for the simple reason that most of that debt was owed to Russia, France, Japan, Germany and other countries. The United States held a minor $2.2 billion of the total debt.

The Paris Club members issued an official press statement:

> The representatives of the Creditor Countries, aware of the exceptional situation of the Republic of Iraq and of its limited repayment capacity over the coming years, agreed on a debt treatment to ensure its long term debt sustainability. To this end, they recommended that their Governments deliver the following exceptional treatment:

- an immediate cancellation of part of the late interest, representing 30% of the debt stock as at January 1, 2005. The remaining debt stock is deferred up to the date of the approval of an IMF standard program. This cancellation results in the write-off of 11.6 billion US dollars on a total debt owed to the Paris Club of 38.9 billion US dollars;
- as soon as a standard IMF programme is approved, a reduction of 30% of the debt stock will be delivered. The remaining debt stock will be rescheduled over a period of 23 years including a grace period of 6 years. This step will reduce the debt stock by another 11.6 billions US dollars increasing the rate of cancellation to 60%;
- Paris Club Creditors agreed to grant an additional tranche of debt reduction representing 20% of the initial stock upon completion of the last IMF Board review of three-years of implementation of standard IMF programs.[21]

The debt relief of Iraq, in which the principal occupier, the United States, generously wrote off the debt owed by Saddam to Washington's rivals who had opposed the war on Iraq—Russia, France, China—was bound with the proviso that Iraq adhere to the strict IMF "standard program." That standard program was the same as the one applied to Indonesia, Poland, Croatia, Serbia, Argentina and post-soviet Russia. It mandated Iraq to turn its economic sovereignty over to IMF technocrats effectively controlled by the US Treasury and the Washington administration.

Adding insult to injury, that old Iraqi debt of the Hussein era was what international governments called "odious debts"—debts incurred without the consent of the population and not in the interest of that population—in short, illegitimate, as the debts of the defunct Soviet Union had been. That did not bother Washington, London and other members of the Paris Club. The debt was a useful weapon to control the "new" Iraq, and to force its transformation into a "free market." GM seeds and the industrialization of agriculture would be at the heart of that forced change.

Privatization of state enterprises was at the top of the Washington Consensus IMF program. Free-market private enterprise was also at the heart of the CPA's 100 Orders of April 2004. This was hardly a coincidence.

The IMF could be accurately labeled as the "policeman of globalization." Since the debt crisis of the 1980's, the IMF enforced brutal creditor austerity and debt repayment plans in developing economies. The debt terms of the IMF were used to force countries to virtually give away their most precious economic assets to foreign interests in order to repay a debt that grew ever larger.

Typically, giant corporate banking and private interests stood behind these IMF measures. They systematically imposed privatization of state enterprises, elimination of public subsidies for food, health and energy, and cuts in public education spending. Every policy that would allow multinational corporations to dominate postwar Iraq would thus be executed by the IMF and the Bremer laws: a shrunken state, flexible workforce, open borders, minimal taxes, no controls on capital outflows out of Iraq, no tariffs, and no ownership restrictions.

The people of Iraq would lose hundreds of thousands of jobs, and foreign products would force domestic Iraqi goods out of the market, of which food would be one major product. Local businesses and family farms would be unable to compete under the imposed rules and foreign competition.

A typical victim of IMF conditions would inevitably be forced to transform its national economy towards export in order to earn dollars to repay their debt. The "carrot" for this was always promise of an IMF "bailout" or "rescue" loan. The blackmail behind the IMF carrot was the threat that a victim debtor country would be permanently blacklisted from all foreign credits should it refuse the IMF conditions.

Iraq was to be no different. The US-mandated Iraqi elections were intended to set the legal stage to bind Iraq's government to the severe IMF controls. In effect, this would place the IMF as the "neutral" agency responsible for Iraqi adherence to the 100 Bremer Orders. The IMF would force Iraq to join Washington's global vision of a "free market."

The IMF planned to reach a specific arrangement with Iraq's new government sometime after the January 30, 2005 Iraqi elections. Since the relief of a large amount of Iraq's external debt was dependent on approval by the Fund, the IMF had considerable leverage in its negotiations with Iraqi leaders.[22]

United Nations Security Council Resolution No. 1483 had given Bremer the power to manage occupied Iraq, but this was to be within the parameters of international law. Bremer's 100 Orders and economic "shock therapy" however, were undertaken in utter violation of international law.

As protests against the Iraqi privatization and violent attacks on American companies spread, it became urgent to conceal this embarrassing fact. Bremer therefore rushed back to Washington to discuss with the President a new scheme for taking over Iraq's economy. The result was the interim regime of Iyad Allawi, and the announcement of Iraqi elections for January 2005. Allawi, a hand-picked Washington protégé who had worked with the CIA for years, was to "legally" implement the illegal Bremer decrees.

Under Order 39 of what became known in Iraq as "the Bremer Laws," Iraqi industries and markets were to be opened to foreign investment with few restrictions. These laws were formulated in a way which would make it very difficult for either the interim government or any subsequent Iraqi government to revoke or repeal these policies.

Indeed, Bremer cemented the 100 Orders with Article 26 of the Iraqi interim constitution, which ensured that once sovereignty was handed over to the interim government, it would be powerless to change the Bremer Laws. In addition, hand-picked US sympathizers were inserted by Bremer into every Iraqi ministry, and were empowered with the authority to override any of the decisions made by subsequent Iraqi governments.

The presence of 132,000 US troops across Iraq, firmly embedded in some 14 new US military bases constructed across the country after 2003, were to guarantee that. It was becoming clear to most Iraqis by late 2004 just what Washington meant when it used the noble words, "planting the seeds of democracy" in their nation.

The seeds had nothing to do with the ability of ordinary Iraqi citizens to determine their own independent destiny.

After official authority was transferred in June 2004 from Bremer's CPA to the Interim Iraqi regime headed by CIA asset, Allawi, the latter agreed to accept debt relief in exchange for its "openness" to IMF-imposed reforms. Thus, in a memorandum attached to a "letter of intent" sent by Central Bank Governor Shababi and Finance Minister Al-Mahdi to the IMF that September, the men expressed their US-installed government's eagerness to "engage" with the fund.[23]

"New financial sector legislation has paved the way for the creation of a modern financial sector," the letter boasted, going on to state that "three foreign banks have already been licensed to begin operations," and that "a number of foreign banks have shown interest in acquiring a minority ownership stake in private Iraqi banks." One bank was the London HSBC, which is among the largest in the world.[24]

The forced transformation of Iraq's food production into patented GMO crops is one of the clearest examples of the manner in which Monsanto and other GMO giants are forcing GMO crops onto an unwilling or unknowing world population.

Notes

1. Coalition Provisional Authority, *CPA Official Documents, Orders,* http://www.cpa-iraq.org/regulations/#Orders.

2. Naomi Klein, "Baghdad Year Zero", *Harpers' Magazine,* September 2004.

3. Coalition Provisional Authority, *op. cit.* In its introduction, the document declares, "Orders—are binding instructions or directives to the Iraqi people that create penal consequences or have a direct bearing on the way Iraqis are regulated, including changes to Iraqi law."

4. Coalition Provisional Authority, *CPA Official Documents, Order 81:Patent, Industrial Design, Undisclosed Information, Integrated Circuits and Plant Variety Law,* http://www.iraqcoalition.org/regulations/index.html#Regulations. See also Focus on the Global South and GRAIN, *Iraq's New Patent Law: A Declaration of War against Farmers,* http://www.grain.org, and Vandana Shiva, *Biopiracy: The Plunder of Nature and Knowledge,* Green Books, Devon, UK, 1998.

5. William Erskine, *Agriculture System in Iraq Destroyed: Self Sufficiency in Food Production Years Away...,* Press Release, 30 June 2003, http://www.icarda.org/News/2003News/30June03.htm.

6. Hope Shand, "Patenting the Planet", *Multinational Monitor,* June 1994, p. 13.

7. GRAIN Press Release, *Iraq's New Patent Law: A Declaration of War Against Farmers,* Focus on the Global South and GRAIN, October 2004, http://www.grain.org/articles/?id=6.

8. *Ibid.*

9. Jeremy Smith, "Iraq: Order 81", *The Ecologist,* February 2005.

10. Portal Iraq, *Seeds for the Future of Iraqi Agriculture,* 27 September 2004, http://www.portaliraq.com/news/Seeds+for+the+future+of+Iraqi+agriculture__529.html.

11. Daniel Stolte, "In the Trenches", *The Business Journal of Phoenix,* 10 June 2005.

12. *Ibid.*

13. Christopher D. Cook, "Agribusiness Eyes Iraq's Fledgling Markets", *In These Times,* http://www.mindfully.org/GE/2005/Iraq-US-Agribusiness-Profit15mar05.htm, 15 March 2005.

14. *Ibid.*

15. IRIN News, *IRAQ: Interview with Minister of Agriculture,* http://www.irinnews.org, Baghdad, 16 December 2004.

16. Ann M. Veneman, *Remarks by Agriculture Secretary Ann M. Veneman to the National Association of Farm Broadcasters Annual Convention,* 14 November 2003, US Department of Agriculture, Washington D.C., Release No. 0384.03.

17. US Congressional Record, Kissinger Associates, Scowcroft, Eagleburger, Stoga, Iraq and BNL, *Statement by Representative Henry B. Gonzalez*, 28 April 1992, US House of Representatives, Page H2694.

18. John King cited in Christopher D. Cook, *op. cit.*

19. *Ibid.*

20. *Ibid.*

21. Paris Club, *Iraq*, http://www.clubdeparis.org, 21 November 2004.

22. International Monetary Fund, *Iraq—Letter of Intent, Memorandum of Economic and Financial Policies, and Technical Memorandum of Understanding*, Baghdad, 24 September 2004.

23. Governor Shababi cited in Brian Dominick, "US Forgives Iraq Debt To Clear Way for IMF Reforms" *NewStandard*, 19 December 2004.

24. *Ibid.*

CHAPTER 11
Planting the "Garden of Earthly Delights"

US Agribusiness Moves to Dominate

The project of making GMO crops the dominant basic crops on the world agricultural market was the creation of a new enforcement institution which would stand above national governments. That new institution, which opened its doors in 1995 was to be called the World Trade Organization (WTO).

In September 1986, two years after the Rockefeller Foundation had launched its genetic engineering rice project, US agribusiness threw its now considerable weight behind a radical new international trade regime, the GATT Uruguay Round.

It was a culmination and a logical consequence of thirty some years of work. The work had begun in the 1950's at Harvard University, under the project financed by the Rockefeller Foundation, designed by Wassily Leontief, and implemented, step-by-step, by Harvard Business School Professors, Ray Goldberg and John Davis, under the slogan of "vertical integration."

After three decades of systematic destruction of barriers to monopoly and vertical integration, of eradication of health regulations and safety standards within the United States agricultural sector, the emerging corporate colossus of agribusiness next moved to flex its muscle by demanding the creation of a new supranational, non-elected body to enforce its private agenda of concentration on a global scale.

The WTO headquarters were established in Geneva, Switzerland, a nominally neutral, scenic, and peaceful location. Behind this façade, however, the WTO was anything but peaceful or neutral. The WTO had been created as a policeman, a global free trade enforcer, and, among its major aims, a battering ram for the trillion dollar annual world agribusiness trade, with the agenda to advance the interests of private agribusiness companies. For that reason, the WTO was designed as a supranational entity, to be above the laws of nations, answerable to no public body beyond its own walls.

GATT agreements had no enforceable sanctions or penalties for violating agreed trade rules. In contrast, the new WTO did have such punitive leverage. It had the power to levy heavy financial penalties or other sanctions on member countries in violation of their rules. The WTO had emerged as a new weapon which could force open various national barriers and which could thereby enhance the proliferation of the soon-to-be commercialized genetically modified crops.

The idea of a WTO, as with most major post-war free trade initiatives, came from Washington. It was the outcome of the GATT Uruguay Round of trade liberalization talks, which began in Punte del Este, Uruguay, in September 1986, and concluded in Marrakesh, Morocco, in April 1994.

Ever since 1948 and the initial founding of the General Agreement on Tariffs and Trade, Washington had fiercely resisted including agriculture into world trade talks, fearing any common international rules would open US markets to foreign food imports and would damage American agriculture's competitiveness. Since the 1950's, US agricultural exports had been a strategic national priority tied to Cold War geopolitics.

Unlike all previous GATT trade rounds, the Uruguay Round made trade in agriculture a main priority. The reason was simple. By the mid-1980's, backed by the aggressive policies of deregulation and free market support from the Reagan Administration, American agribusiness was powerful enough to launch its global trade offensive, and in a big way.

The Washington position on the Uruguay Round agricultural agenda had been drafted by Cargill Corporation of Minneapolis, Minnesota. Daniel Amstutz, a former Cargill executive and Special Ambassador for the Reagan Administration at the GATT, drew up the four-point Amstutz Plan.[1]

It was, in fact, the Cargill Plan. Cargill was then the dominant US private agribusiness giant, with global sales well over $56 billion, and plants in 66 countries around the world. It had built its mighty global empire through working with the Rockefeller interests in Latin America, as well as with Henry Kissinger in the 1970's Great Grain Robbery sales of US wheat to the Soviet Union at huge profit. Its influence on Washington, and especially on US Department of Agriculture policy, was immense.[2]

The four Amstutz demands at the GATT talks worked uniquely to the benefit of US agribusiness and their growing global position. The points included a ban on all government farm programs and price supports worldwide; a prohibition on countries who seek to impose import controls to defend their national agriculture production; a ban on all government export controls of agriculture, even in time of famine. Cargill wanted to control world grain export trade.

The final Amstutz demand, presented to the GATT Uruguay Round participants in July 1987, implied that GATT trade rules *limit* the right of countries to enforce strict food safety laws! The global "free market" was seemingly more sacred to Cargill and their agribusiness allies than mere human life. National food safety laws were seen by US agribusiness as a major barrier to unfettered pursuit of high profit from low-wage, low quality industrial farm operations in developing countries, as well as in the USA. Furthermore, agribusiness wanted an unrestricted ability to market the new genetically engineered crops, with no national concerns about health and safety getting in their way.

Amstutz was a dedicated champion of agribusiness interests and was named the special liaison of the Bush Administration Department of Agriculture to Iraq in 2003 to direct the transformation of Iraqi farming into US-led, "market oriented" export agribusiness with GMO crops, described in chapter 10.

The main US agriculture demands at the Uruguay Round centered on the call for a mandatory end to state agriculture export subsidies, a move aimed squarely at the European Community's Common Agriculture Program (CAP). Washington called the process "agriculture trade liberalization." The beneficiaries were US agribusiness, i.e. the dominant players, in a scenario reminiscent of how British free trade demands in the late 1870's served the interests of British international business and banking, then the dominant world players.

The IPC and the Agribusiness Lobby
Cargill was one of the main drivers of the US Business Roundtable, a powerful lobby made up of the largest US corporate executives. The Business Roundtable formed an alliance for GATT in 1994 to lobby the US Congress to accept its agriculture agenda—which it did almost without question.

The Congressional decision to back the GATT and the creation of the new WTO was made easier by the fact that Cargill and their Business Roundtable friends poured millions of dollars in campaign contributions to support key members of the US Congress.[3]

Not content to put all its eggs in one basket, Cargill also created the Consumers for World Trade (CWT), another "pro-GATT" lobby which, curiously enough, represented not consumers but agribusiness and multinational interests. Corporate membership cost $65,000. Cargill also formed an Emergency Committee for American Trade, to convince Congress to accept the new WTO agriculture agenda.[4]

The international lobby working with Cargill and US agribusiness to push through the radical GATT agriculture agenda was an obscure and powerful organization, which called itself the International Food and Agricultural Trade Policy Council, or IPC.

Founded in 1987 to promote the liberalization of agricultural trade and in particular, the Amstutz Plan for agribusiness, the IPC included top executives and officials from Cargill, GMO giant Syngenta (then Novartis), Nestlé, Kraft Foods, Monsanto, Archer Daniels Midlands (ADM), Bunge Ltd., Winthrop Rockefeller's Winrock International Foundation, the US Department of Agriculture and Japan's largest trading group, Mitsui & Co. IPC was an interest group few politicians in Brussels, Paris or elsewhere could afford to ignore.

Cargill, the IPC, and the Business Roundtable all worked closely with Clinton Administration US Trade Representative and later Commerce Secretary, Mickey Kantor. By presenting the WTO as being substantially similar to GATT consensus rules—thus, by essentially lying—Kantor got the Uruguay Round's WTO proposal through the US Congress.

WTO rules were to be dominated by a Group of Four, the so-called QUAD countries—the USA, Canada, Japan and the EU. They could meet behind closed doors and decide policy for all 134 nations. Within the QUAD, the US-led agribusiness giants controlled major policy. In effect it was a consensus, but a consensus of private agribusiness, that determined WTO policy.

The WTO Agreement on Agriculture, which was written by Cargill, ADM, DuPont, Nestlé, Unilever, Monsanto and other agribusiness corporations, was explicitly designed to allow the destruction of national laws and safeguards against the powerful pricing power of the agribusiness giants.

By 1994, as the WTO was in the process of being established, it had become Washington policy to give full backing to the development of genetically-modified plants as a major US strategic priority. The Clinton Administration had made "biotechnology," along with the Internet, a strategic priority for US Government backing, formal as well as informal. Clinton gave full support to Mickey Kantor as chief negotiator for the WTO ratification process.

When Kantor left the Washington Government in 2001, he was rewarded for his service to US agribusiness during the GATT negotiations. Monsanto Company, then the world's most aggressive promoter of genetically modified crops and related herbicides, named Kantor to be a member of Monsanto's Board of Directors. The revolving door between the government and the private sector was well oiled.

Monsanto, DuPont, Dow Chemical and other agricultural chemical giants had transformed themselves into controllers of patented genetically-modified seeds for the world's major staple crops. The time was ripe to establish a police agency which could force the new GMO crops on a skeptical world. The WTO Agreement on Agriculture was to be the vehicle for that, along with the WTO rules enforcing Trade Related Intellectual Property Rights (TRIPS).

WTO and Bad TRIPS

The WTO marked a step for the globalization of world agriculture, under terms defined by US agribusiness. WTO rules would open the legal and political path to the creation of a global "market" in food commodities similar to that created by the oil cartel under the Rockefeller Standard Oil group a century before. Never before the advent of agribusiness had agriculture crops been viewed as a pure commodity with a global market price. Crops had always been local along with their markets, the basis of human existence and of national economic security.

Washington's Amstutz Plan, with slight modifications, became the heart of the WTO's Agreement on Agriculture, or AoA as it came to be known. The policy goal of the AoA was to create what agribusiness deemed its highest priority—a free and integrated global market for its products. While talking rhetorically about "food security," it mandated that such security would only be possible under a regime of free trade, an agenda uniquely beneficial to the giant global grain traders such as Cargill, Bunge and ADM.

In 1992, the senior Bush Administration made the ruling, without public debate, that genetically engineered or modified food or plants were "substantially equivalent" to ordinary seeds and crops and hence needed no special government regulation. That principle was enshrined into the WTO rules under its "Sanitary and Phytosanitary Agreement," or the SPS. Phytosanitary was a fancy scientific term which simply meant that it dealt with plant sanitation, i.e. with GMO plant issues.

The crafty formulation of the SPS rule stipulated that "food standards and measures aimed at protecting people from pests or animals *can be potentially* used as a deliberate barrier to trade," and hence, must be forbidden under WTO rules.[5] Under the guise of appearing to enshrine plant and human health safety into WTO standards, the IPC and the powerful GMO interests within it ensured just the opposite. Few politicians in WTO member countries bothered to even read past the formidable term, "phytosanitary." They listened to their own agribusiness lobby and approved.

Under the WTO's SPS rule national laws banning genetically modified organisms from the human food chain, because of national health concerns regarding potential threat to human or animal life, were termed "unfair trade practices."[6] Other WTO rules prohibited national laws that required labeling of genetically engineered foods, declaring them to be "Technical Barriers to Trade."[7] Under the WTO, "trade" was deemed a higher concern than the citizen's right to know what he or she was eating. Which trade and to whose benefit was left unspoken.

Parallel to the international negotiations that ultimately created the WTO, some 175 nations were negotiating safeguards to insure that biological diversity and food safety concerns remained a priority in face of the onslaught of new, largely untested GMO crops.

In 1992, two years before the final WTO document was agreed upon, the 175 participating nations had signed a UN Convention on Biological Diversity (CBD). That CBD dealt with the safe transfer and use of GMOs.

As an extension of that convention, numerous governments, especially in developing countries, felt that a protocol explicitly dealing with the potential risks of GMOs was necessary. At that point, GMOs were still largely in a testing phase.

Despite strong resistance, especially from the US Government, a formal working group began to draft a Biosafety Protocol in 1996. Finally, after seven years of intense international negotiation involving hearings from relevant interest groups from around the world,

138 member nations of the UN met in Cartagena, Colombia, expecting to sign the final Biosafety Protocol to the UN Convention on Biological Diversity.

They were too optimistic. The demands of developing countries including Brazil and several African and Asian nations were ambushed by the powerful organized government and agribusiness lobby backing GMO. After ten days of non-stop debate, delegates were stymied by opposition from pro-GMO countries. Canada, acting as spokesman for what was called the Miami Group, led by the United States and other pro-GMO agribusiness countries, won an agreement to adjourn without an agreement and continue work in a smaller committee. The talks had been sidetracked by the Miami Group, six countries led by the USA and including Canada, a close backer of US GMO policy; Argentina, by then fully in the grip of Monsanto and US agribusiness; Australia, another agribusiness free trade ally of Washington; Uruguay and Chile, two countries whose ties to Washington were extremely close. Curiously, the United States Government was not officially present at the Cartagena meetings. The Clinton Administration, an ardent GMO supporter, had refused to participate as they had refused to sign the earlier Convention on Biological Diversity.

Unofficially, however, Washington representatives orchestrated the entire Miami Group sabotage of the talks. The demands of the Miami Group were simple. They insisted that WTO trade rules be formally written into the protocol and that it be stated that biosafety measures must remain subordinate to WTO trade demands. Their argument was insidious and sophistic. They turned the tables and argued, not that the safety of GMO crops was unproven, but rather that the biosafety concerns of most member countries to the Convention over risks of GMO were unproven and hence, should be considered a "barrier to trade."[8] In such a case, Miami Group countries insisted WTO rules prohibiting unfair trade barriers must take precedence over the Biosafety Protocol.

The talks collapsed. Little was heard again of the Cartagena Biosafety Protocol. Washington, the WTO and the GMO interests behind them had cleared the path to the unfettered spread of GMO seeds worldwide.

The doctrine of the WTO was simple: free trade—on terms defined by giant private agribusiness conglomerates—was to reign supreme above sovereign nation states and above the concern for human or animal health and safety. "Free market über Alles" was the motto.

Having Your Cake and Eating it too

Washington argued that only products which had been "substantially transformed" could be labelled in a given country. They then asserted under the Bush 1992 ruling that their genetically modified crops were "substantially equivalent" to ordinary plants, not "substantially transformed," and hence, needed no label.

Yet, US patent law allowed agribusiness companies at the same time to claim exclusive patent rights on their genetically modified organisms or seeds, on the argument that the introduction of a foreign DNA into the genome of a plant such as rice uniquely altered the plant or, one could say, "substantialy transformed" it.

The contradictions between the "substantially equivalent" ruling of Washington on GMO products and allowing radical new patents on GM seeds to be deemed "substantially transformed" did not bother many Washington officials. Whatever argument it took to advance the agribusiness gene revolution agenda was fine by them. Niceties of logical consistency weren't high on Washington's list of priorities in promoting their gene revolution.

The legal framework for patenting plants was enshrined in WTO rules protecting Trade Related Intellectual Property Rights (TRIPS). Under TRIPS, all member nations of WTO were required to pass laws to protect patents (intellectual property rights) for plants. The patents would block anyone but the patent holder from making, selling or using the "invention." This little-noticed proviso in the new WTO rules opened the floodgates for US and international agribusiness to advance the Rockefeller Foundation genetic engineering strategic agenda.

The WTO TRIPS rules permitted well-financed agrochemical multinationals with large R&D budgets to set the stage for demanding royalty payments or even denying a customer or country its patented seed. In the case of plants, the exclusive patent was in force for twenty years. As one critical scientist put it, under TRIPS and genetic patent law, "knowledge is property. It belongs to corporations and is not accessible to farmers."[9]

Backed by the police powers of the WTO and the muscle of the US State Department, the gene multinationals—Monsanto, Syngenta and others—soon began testing the limits of how far they could impose the patenting of plants and other life forms on other countries.

A Texas biotechnology company, RiceTec, decided it would take out a patent on Basmati rice, the variety which has been the dietary staple in large parts of India, Pakistan, and Asia for thousands of years. In 1998, RiceTec took a patent on its genetically modified Basmati rice, and thanks to US laws forbidding the labelling of genetic foods RiceTec was able to sell it legally by labelling it as ordinary Basmati rice. RiceTec, it turned out, had gotten a hold of the precious Basmati seed, which had been put in trust by dubious means at the Rockefeller Foundation's International Rice Research Institute in the Philippines.[10]

The IRRI had made a "safety" duplicate of the invaluable collection of rice seeds collected in the Philippines and stored it in a seed bank at Fort Collins, Colorado, making highly dubious the claim by IRRI that the seeds would be stored as a secure seed resource for the region's rice farmers. IRRI had convinced rice farmers that giving their invaluable seed varieties to the IRRI was for their own security.

In Colorado, far from the Philippines, the IRRI gave the valuable seeds to RiceTec scientists, who then patented it. They knew it was highly illegal; even in Texas, rice scientists know that Basmati rice doesn't grow normally on the dusty plains around Crawford, Texas.[11]

RiceTec, with the collusion of IRRI, stole the seeds for its patent. Yet under the carefully crafted rules set up by the Rockefeller

Foundation's IRRI, while seeds from the gene bank should not be patented, once a scientist manages to do breeding work, regardless of how they may patent it.

In December 2001, the US Supreme Court enshrined the principle of allowing patents on plant forms and other forms of life in a groundbreaking case entitled J.E.M. Ag Supply vs. Pioneer Hi-Bred. The United States Supreme Court granted certiorari to determine whether newly developed plant breeds fall within the subject matter of 35 U.S.C. § 101, or whether the alternative statutory regimes provided by Congress showed a legislative intent that the regular utility patent statute not cover plants. To the surprise of most legal experts, the Court ruled GMO plant breeds could be patented.[12]

From that point onwards, the genetic agribusiness cartel had the backing of the highest court in the United States. This could now be used as a battering ram to force other, less powerful countries to respect US GMO seed patents.

The complicity of essential US Government agencies, legally and nominally responsible for ensuring public health and safety of the general population, was a decisive part of the GMO revolution. In a full-page *New York Times* exposé that ran on January 25, 2001, the paper wrote that Monsanto gained "astonishing" control over its own regulatory industry, through the Environmental Protection Agency, the Department of Agriculture and the Food and Drug Administration. Dr. Henry Miller, who was in charge of biotechnology issues for the Food and Drug Administration from 1979 to 1994, told the *Times*: "In this area, the U.S. government agencies have done exactly what big agribusiness has asked them to do and told them to do."

Ironically, Monsanto, Syngenta, DuPont and the other major holders of patents on genetically modified plants claimed that genetically engineered rice, corn, soybeans and other crops would solve the problem of world hunger and lead to greater food security. In fact, their aggressive patenting of plant varieties led to restricted research, reduced genetic plant diversity, and concentrated ownership of seeds which had been for thousands of years the heritage of mankind. This process enormously increased the risk for entire plant species to be devastated due to the new monocultures.

The Four Horsemen of the GMO Apocalypse
With the full backing of the powerful WTO, and the USA and UK governments, the major international biotech companies consolidated their grip, using genetically modified patents on every plant imaginable. The Gene Revolution was a monsoon force in world agriculture by the end of the 1990's.

By 2004, four global private companies dominated the market for genetically modified seeds and their related agrichemicals. The world's number one GMO company was the Monsanto Corporation of St. Louis, Missouri, the leading provider of genetically modified seeds and the world's largest producer of the chemical herbicide, glyphosate, which it called its Roundup group of herbicides. Beginning in the 1990's, Monsanto spent some $8 billion buying up seed companies to complement its role as one of the world's leading herbicide producers.

The strategy defined in an interview in the April 12, 1999 *Business Week* of Monsanto CEO, Robert B. Shapiro, was to create a global fusion of "three of the largest industries in the world—agriculture, food and health—that now operate as separate businesses. But there are a set of changes that will lead to their integration."[13] Monsanto saw itself as a kind of modern day King Kanute, moving that sea of changes on command.

Monsanto was founded in 1901 to manufacture industrial chemicals such as sulphuric acid. It produced and licensed most of the world's polychlorinated biphenyls, which later proved to cause severe brain damage, birth defects and cancer.

In early 2007, British investigators uncovered internal UK Government memos and evidence that Monsanto had illegally dumped some 67 chemicals, including Agent Orange derivatives, dioxins and PCBs (which could have been made only by Monsanto), from one unlined porous quarry in South Wales that was not authorised to take chemical wastes, polluting underground water supplies and the atmosphere 30 years later. *The Guardian* reported,

"Evidence has emerged that the Monsanto chemical company paid contractors to dump thousands of tonnes of highly toxic waste in British landfill sites, knowing that their chemicals were liable to contaminate wildlife and people."[14] Monsanto entered the world of GMO with a less than spotless record for corporate integrity, or demonstrated concern for human health.[15]

The second member of the GMO global quartet to emerge in the late 1990's was DuPont Corporation's Pioneer Hi-Bred International, Inc., of Johnstown, Iowa. Pioneer Hi-Bred billed itself as "the world's leading developer and supplier of advanced plant genetics to farmers worldwide," and was active in 70 countries.

Pioneer Hi-Bred, a company founded in the 1930's by later Rockefeller collaborator in the Green Revolution, Henry Wallace, was taken over by the Delaware chemicals giant DuPont in 1999. With its huge germplasm holdings and intellectual property, Pioneer Hi-Bred International (PHI) was considered to be the largest proprietary seedbank in the world. Pioneer's market dominance was based primarily on its corn seed.

Since the 1980s, Pioneer had been moving into plant genetics. In October 1999, DuPont completed its $7.7 billion takeover of Pioneer, and created a seed-chemical industry complex intended to be a primary engine in the shift by the chemical industry from petroleum-dependency, to a feedstock provided by genetic engineering.[16]

Based in Indianapolis, Indiana, the third GMO giant was Dow AgroSciences, a $3.4 billion seed and agrochemicals conglomerate active in 66 countries. Dow AgroSciences was formed in 1997 when Dow Chemical bought drug-maker Eli Lilly's stake in Dow Elanco. The parent company had grown to be Dow Chemical, the world's second largest chemical company overall, with annual revenues of over $24 billion and operations in 168 countries.[17]

Like its GMO agribusiness allies, Monsanto and DuPont, Dow had a disreputable history concerning environmental and public health issues. Dow's factories at its global headquarters in Midland, MI, contaminated the entire region, including the Tittabawassee River floodplains, with stratospheric levels of dioxin. Tests done by the Michigan Department of Environmental Quality found that 29 of 34 soil samples taken in Midland had dioxin levels higher than state cleanup standards.[18] Some samples had concentrations of dioxin nearly 100 times higher than cleanup standards. The state warned residents of Midland to "avoid allowing children to play in soils. Wash hands and any other exposed body surfaces after any soil contact. Do not eat unwashed foods from your garden. Do not engage in any other activities that may introduce soil into the mouth...."[19]

Dioxin is among the most toxic compounds ever studied. It is harmful to life in miniscule amounts and has been linked by experts to endometriosis, immune system impairment, diabetes, neurotoxicity, birth defects, decreased fertility, testicular atrophy, reproductive dysfunction, and cancer. Dioxin can affect insulin, thyroid and steroid hormones, threatening the development of all human newborns, according to one scientific report.[20]

Dow was the innovator of the infamous napalm used against civilians in Vietnam. The jelly-like chemical, when sprayed over people, would burn them on contact. The infamous 1972 photograph of a naked child running down a street in Vietnam screaming in agony, captured for the world the effects of napalm. Dow's President at the time, Herbert D. Doan, described napalm as "a good weapon for saving lives ... a strategic weapon essential to the pursuit of the tactic we are engaged in without exorbitant loss of American lives."[21]

Dow AgroSciences described its GMO role as "providing innovative crop protection, seeds, and biotechnology solutions to serve the world's growing population." In 2003, in the case Bates v. Dow AgroSciences, twenty-nine farmers in western Texas went to court contending that *Strongarm*, a herbicide manufactured by Dow AgroSciences, had severely damaged their peanut crops and had not killed the weeds as promised. The farmers sued Dow for false advertising, breach of warranty and fraudulent trade practices under the Texas Deceptive Trade Practices Act. Dow AgroSciences won a declaratory judgment against the farmers in federal district court seeking, among other things, a judicial declaration that the

Federal Insecticide, Fungicide and Rodenticide Act (FIFRA) pre-empted the farmers' state law claims. The US Government sided with Dow as *amicus curiae* in the case, which went to the Supreme Court.[22]

The fourth horseman of the GMO battalion was Syngenta of Basel, Switzerland, which grew from the 2000 merger of the agriculture divisions of Novartis and AstraZeneca into a $6.8 billion agriculture and chemicals company. It claimed in 2005 to be the world's largest agrochemical corporation and third largest seed company. Though it was Swiss-based, Syngenta was in many respects a British-controlled company whose chairman and many directors came from the British AstraZeneca side. Syngenta, which deliberately cultivated a low profile to avoid the controversies plaguing its US rivals, was the world's second biggest agrochemicals producer and third biggest seed producer.

Syngenta became subject of major unwanted media attention in 2004 when a German farmer, Gottfried Glöckner of North Hessen, found evidence that planting Syngenta Bt-176 genetically-engineered corn to feed his cattle in 1997, had been responsible for killing off his cattle, destroying his milk production and poisoning his farmland. Syngenta's Bt-176 corn had been engineered to produce a toxin of Bacillus thuringiensis which they claimed was deadly to a damaging insect, the European Corn Borer.[23]

Glöckner was the first approved farmer in Germany to use the Bt Corn from Syngenta for animal feed. He kept detailed notes of his experiences, initially believing he was at the start of a revolution in agriculture through GMO. In the end, his proved to be one of the longest-running tests of the effects of Syngenta's Bt-176 corn anywhere in the world, lasting almost five years. The results were not encouraging for the proponents of GMO.

A test of GMO's efficacy wasn't Glöckner's intent, however. He wanted the beneficial effects of the GMO crop to feed his cattle, and to eliminate crop loss from the European Corn-borer insect, which typically cut harvest yields up to 20%.

In the first year, 1997, Glöckner was careful. He grew only a small test field of Bt-176 corn. The results were impressive: corn of uniform height, green shoots, "standing tall like soldiers," he recalled. "I was fascinated, as a practitioner, to see high yields, and apparently healthy plants, with no sign at all of damage from the corn-borer." The second year, 1998, he expanded to 5 hectares of Syngenta's Bt-176 corn, working closely with the company's German representative, Hans-Theo Jachmann. By 2000, Glöckner had expanded the GMO experiment to his entire field of some 10 hectares, about 25 acres. Each successive harvest he gradually increased the amount of Bt corn he fed his cattle, carefully noting milk yields, and possible side effects. The first three years no side effects from the rising GMO feedings could be noted.[24]

However when he increased the dosage to a diet of pure GMO corn from his Syngenta green fields, convinced he would gain higher milk yields, he witnessed a nightmare unfold.

Glöckner, a university-trained farmer, told an Austrian journalist that he was shocked to find his cattle having gluey-white feces and violent diarrhea. Their milk contained blood, something unheard of in the midst of lactation. Some cows suddenly stopped producing milk. Then, five calves died, one after the other, between May and August 2001, an extremely alarming event. Glöckner ultimately lost almost his entire herd of 70 cows. Syngenta rejected any responsibility for the events, insisting that the cows would detoxify the toxin of Bacillus thuringiensis in the Bt-176 corn, according to their tests.

Despite Syngenta's denial of any responsibility, Glöckner persisted and obtained independent scientific analyses of his soil, his silo corn and his cows. One lab returned the result that confirmed Glöckner's conviction that Syngenta's Bt-176 GMO corn was the cause. It showed that in his Bt-176 corn from the year 2000, were 8.3 micrograms of toxin per kilogram. In June 2004, Prof. Angelika Hilbeck of the respected Swiss Federal Institute of Technology Geobotanical Institute found that from Glöckner's Bt-176 corn samples Bt toxins were "found in active form and extremely stable," an alarming result that Syngenta insisted was simply not possible.[25] Glöckner's independent test results were in complete contradiction to Syngenta's claims that their Research Center in North Carolina "discovered no Bt toxins in the feed sample."[26]

In 2005, the same Syngenta made a bold move to lock up a major share of GMO Terminator patents. Syngenta applied for patents that could effectively allow the company to monopolize key gene sequences that are vital for rice breeding as well as dozens of other plant species. Syngenta's enthusiasm for the rice genome stemmed from rice's major genetic similarities (i.e., DNA or protein sequences) to other species ranging from maize and wheat to bananas, genetic similarities called "homologies." While Syngenta was donating rice germplasm and information to public researchers with one hand, it was attempting to monopolize rice resources with the other.

Syngenta's controversial relationship with rice and patents included its involvement with GMO Golden Rice and the Syngenta Foundation's membership in the Consultative Group on International Agricultural Research (CGIAR).[27]

GMO and Pentagon Deals

Quite notable was the fact that three of the four global players in GMO were not only American-based, but had decades-long involvement with the Pentagon in supplying war chemicals, including napalm and the notorious Agent Orange plant defoliant used by the US military in Vietnam.

In early 2001, a New Zealand magazine, *Investigate*, reported an alarming discovery. In an article titled, "Dow Chemical's Nasty Little Secret—Agent Orange Dump found under New Zealand Town," a former top official at New Plymouth's Ivon Watkins Dow chemical factory confirmed the worst fears of residents: Part of the town was sitting on a secret toxic waste dump containing the deadly Vietnam War defoliant Agent Orange. "We've buried it under New Plymouth," the official admitted. The article added, "And if any further proof were needed that surplus Agent Orange had been dumped at New Plymouth, local residents found a drum of the chemical on the beach near Waireka Stream." Dow Chemical had kept it secret for 20 years.[28] Civilian and retired military victim lawsuits against the US Government for diseases contracted in Vietnam as a result of exposure to Agent Orange were still being litigated in US courts more than three decades after the end of the Vietnam War.

In 1990, retired Admiral Elmo R. Zumwalt, was named to carry out an investigation of the government's knowledge of the toxicity of Agent Orange to its own soldiers and civilians. Zumwalt's report stated: "From 1962 to 1970, the US military sprayed 72 million liters of herbicides, mostly Agent Orange, in Vietnam. Over one million Vietnamese were exposed to the spraying, as well as over 100,000 Americans and allied troops." Dr. James Clary, a scientist at the Chemical Weapons Branch, Eglin Air Force Base, who designed the herbicide spray tank and wrote a 1979 report on Operation Ranch Hand (the name of the spraying program), told Senator Daschle in 1988:

> When we (military scientists) initiated the herbicide program in the 1960s, we were aware of the potential for damage due to dioxin contamination in the herbicide. We were even aware that the "military" formulation had a higher dioxin concentration than the "civilian" version due to the lower cost and speed of manufacture. However, because the material was to be used on the "enemy," none of us were overly concerned. We never considered a scenario in which our own personnel would become contaminated with the herbicide.[29]

By 2005, the three US leaders in spreading genetically-engineered agricultural seeds and herbicides had built their argument against any government regulation of their research or the safety of its genetic engineered seeds, by claiming that simply trusting them was the most reliable and efficient way of policing GMO safety concerns.

The history of one of the three US makers of Agent Orange, Monsanto, reveals how high that company set the bar for integrity and human life. Keith Parkins described Monsanto's record in Vietnam:

> Monsanto was the main supplier. The Agent Orange produced by Monsanto had dioxin levels many times higher than that produced

by Dow Chemicals, the other major supplier of Agent Orange to Vietnam. Dioxins are one of the most toxic chemicals known to man. Permissible levels are measured in parts per trillion, the ideal level is zero. The Agent Orange manufactured by Monsanto contained 2,3,7,8-tetrachlordibenzo-para-dioxin (TCDD), extremely deadly even when measured against other dioxins. The levels found in domestic 2,4,5-T were around 0.05 ppm, that shipped to Vietnam peaked at 50 ppm, i.e., 1,000 times higher than the norm.

Monsanto's involvement with the production of dioxin contaminated 2,4,5-T dates back to the late 1940s. Almost immediately workers started getting sick with skin rashes, inexplicable pains in the limbs, joints and other parts of the body, weakness, irritability, nervousness and loss of libido .... Internal Monsanto memos show that Monsanto knew of the problems but covered it up.

Parkins concluded that,

> A wide range of products manufactured by Monsanto have been contaminated with dioxins, including the widely used household disinfectant Lysol. Monsanto's attempts at a cover-up were revealed when a court awarded $16 million in punitive damages against Monsanto. It was revealed that Monsanto had intimidated employees to keep quiet, had tampered with evidence, had submitted false data and samples to EPA. An investigation by Cate Jenkins of the EPA Regulatory Development Branch documented a track record of systematic criminal fraud.[30]

An estimated 50,000 Vietnamese children had been born with "horrific deformities" in the regions sprayed with Agent Orange, a practice which stopped only in 1971.[31] It was an operation enormously profitable for Monsanto's chemicals division sales at the time.[32]

In 1999, Canadian national radio, CBC, aired an interview with Dr. Cate Jenkins, an environmental chemist with the US government's Environmental Protection Agency, EPA. Referring to the situation when Monsanto faced lawsuits from US veterans over alleged dioxin poisoning from exposure to Agent Orange, she noted that:

Monsanto was very worried about the impact of being sued by Vietnam veterans. So they were worried about lawsuits. They published a press release during the suit by Vietnam veterans saying our studies show that dioxin does not cause any cancers in humans.

The studies were paid for by Monsanto. The bottom line is that the Vietnam veterans were denied compensation for their cancers, their birth defected children. You could not win a court case when you sued a chemical company for exposures to dioxin ... I am a chemist, an environmental scientist working for the Environmental Protection Agency since 1979. I was able to examine the actual statements of the scientist who had conducted the studies for Monsanto. And those were quite revealing. My evaluation of the studies, I would use the word, rigged. They designed a study to get the results that they wanted. The unexposed population that was supposed to be dioxin free actually did have exposures. Also certain key cases of cancers were eliminated from the Monsanto study for spurious reasons.[33]

Jenkins was transferred to another EPA department and harassed for more than two years as a result of going public.

In 1984, Monsanto, Dow Chemicals and the other makers of Agent Orange paid $180 million into a fund for US military veterans following a bitter, long lawsuit. They refused to admit wrongdoing. More than a decade later, the same companies had refused to pay a single cent to the Vietnamese victims of Agent Orange poisoning.

In 2004, President George W. Bush's Administration cancelled an agreed US-Vietnamese project to examine the long-term genetic impact of Agent Orange. Agent Orange was hardly the theme Monsanto wanted the world public to associate with the greatest global supplier of genetically modified food crops, crops it claimed were designed to feed the hungry of the world. Unlike some politically correct politicians, Monsanto was not into public apologies for its actions.

Letting the GMO Genie Out of the Bottle
By the middle of the 1990's, with the backing of the WTO and Washington, those same gene giants—Monsanto, Dow, DuPont,

Syngenta, and a small handful of others—turned their patented seeds loose on the world.

In 1996, Monsanto shipped to Europe a container full of soybeans from the US. It was not labeled, and EU inspectors only later discovered that it contained Monsanto genetically modified soybeans, the same soybeans which they had spread across Argentina. It entered the food chain without labeling. The EU replied with a moratorium on the commercialization of GM crops in late 1997.[34]

When George W. Bush made the proliferation of GM seeds highest priority after the Iraq war in 2003, the seed cartel led by Monsanto had already spread its patented seeds with alarming speed. Bush's prime goal was to force the lifting of the 1997 EU ban on GM seed commercialization in order to open the next major markets for GMO takeover.

By 2004, according to a report from the Rockefeller Foundation-funded International Service for the Acquisition of Agri-biotech Applications (ISAAA), the planting of genetically engineered crops worldwide had grown by an impressive 20% compared with a year earlier. If it was the ninth such double digit increase since 1996, and the second highest on record. More than 8 million farmers in 17 countries planted GMO crops, and 90% of those were from poor developing countries, precisely the aim of the original Rockefeller Foundation gene revolution.[35] Following the United States as world GMO crop leader, Argentina, Canada and Brazil were by far the largest genetically engineered food producers worldwide.

The ISAAA also noted that GMO soybeans made up 56% of all soybeans planted in the world; GMO corn made up 14% of all corn, GMO cotton was 28% of world cotton harvest, and GMO canola, a form of rapeseed oil, totalled 19% of all world rapeseed harvest.[36] Canola oil, toxic in the human diet, was developed as a genetically modified product in Canada, where, in a burst of marketing patriotism, it was labelled Canadian oil or Canola.[37]

In the United States, with aggressive Government promotion, absence of labelling, and the domination of US farm production by agribusiness, genetically engineered crops had essentially taken over the American food chain. In 2004, more than 85% of all US soybeans planted were genetically modified crops, and most were from Monsanto. 45% of all US corn harvested was GMO corn.[38] Corn and soybeans constituted the most important animal feed in US agriculture, which meant that nearly the entire meat production of the nation as well as its meat exports had been fed on genetically modified animal feed. Few Americans had a clue as to what they were eating. No one bothered to tell them, least of all the Government agencies entrusted with a mandate to protect citizens' health and welfare.

The spread of large fields dedicated to planting GMO crops led to contamination of adjacent organic crops to the point that after just six years, an estimated 67% of all US farm acreage had been contaminated with genetically engineered seeds.

The genie was out of the bottle.

It was not a process which could be reversed in any way known to science. A 136 page review of all known worldwide studies of GMO effects, prepared by an internationally respected group of scientists led by Dr. Mae, presented sobering thoughts about the advisability of untested release of GMO plants into world agriculture. The study warned:

> The most obvious question on safety is with regard to the transgene and its product introduced into GM crops, as they are new to the ecosystem and to the food chain of animals and human beings.[39]
> The Bt toxins from Bacillus thuringiensis, incorporated in food and non-food crops, account for about 25% of all GM crops currently grown worldwide. It was found to be harmful to mice, butterflies and lacewings up the food chain. Bt toxins also act against insects in the Order of Coleoptera (beetles, weevils and styloplids), which contains some 28,600 species, far more than any other Order. Bt plants exude the toxin through the roots into the soil, with potentially large impacts on soil ecology and fertility.[40]

The group of scientists, which included Dr. Arpad Pusztai, concluded from their investigation that:

Bt toxins may be actual and potential allergens for human beings. Some field workers exposed to Bt spray experienced allergic skin sensitization and produced IgE and IgG antibodies. A team of scientists has cautioned against releasing Bt crops for human use. They demonstrated that recombinant Cry1Ac protoxin from Bt is a potent systemic and mucosal immunogen, as potent as cholera toxin. A Bt strain that caused severe human necrosis (tissue death) killed mice within 8 hours, from clinical toxic-shock syndrome. Both Bt protein and Bt potato harmed mice in feeding experiments, damaging their ileum (part of the small intestine). The mice showed abnormal mitochondria, with signs of degeneration and disrupted microvilli (microscopic projections on the cell surface) at the surface lining the gut."[41]

The Independent Science Panel report stated that in this regard:

Because Bt or Bacillus thuringiensis and Bacillus anthracis (anthrax species used in biological weapons) are closely related to each other and to a third bacterium, Bacillus cereus, a common soil bacterium that causes food poisoning, they can readily exchange plasmids (circular DNA molecules containing genetic origins of replication that allow replication independent of the chromosome) carrying toxin genes. If B. anthracis picked up Bt genes from Bt crops by horizontal gene transfer, new strains of B. anthracis with unpredictable properties could arise.[42]

Licensing Life Forms

Just as they vigorously insured a non regulatory regime, the GMO seed cartel imposed rigid licensing and technology agreements insuring annual royalties to Monsanto and the other biotech seed companies from farmers using their seeds. The private companies were not at all anti-government; they only wanted Government rules to serve their private interests.

As with other gene seed companies, Monsanto required farmers to sign a Technology Use Agreement which tied them to paying fees each year to Monsanto for its "technology," namely, genetically engineered seeds.

As independent seed suppliers were rapidly being swallowed up by Monsanto, DuPont, Dow, Syngenta, Cargill or other large agribusiness firms, farmers increasingly were trapped into dependency on Monsanto or the other GMO seed suppliers. American farmers were among the first to experience this new form of serfdom.

With the 2001 US Supreme Court ruling, GMO firms like Monsanto could intimidate US farmers into becoming "seed serfs." The Monsanto penalty for not paying the fees was severe punitive legal damages in a court trial. Monsanto also made certain it would have a friendly court hearing. It had written into its master contract the provision that any litigation against the company be heard in St. Louis, where jurors knew that Monsanto was a major local employer.

Monsanto and the other GMO seed companies demanded that farmers pay each year for new seeds. The farmers were forbidden to re-use seeds from the previous years. Monsanto went so far as to hire private Pinkerton detectives to spy on farmers to see if they were reusing their old seeds. In some areas in the US, the company advertised free leather jackets to anyone informing on a farmer using old Monsanto seed.[43]

Notably, the big four major suppliers of genetically engineered agriculture seed—Monsanto, Syngenta, Dow and DuPont—had originated as and still were major chemical companies. The reason was the same in each case. They all originally manufactured pesticide and herbicide chemicals before they ever ventured into genetic engineering of seeds.

During the early 1990's, the herbicide giants had reorganized themselves as "life sciences" companies. They bought up the existing seed companies, large and small. They forged alliances with transporters and food processors, and emerged at the heart of the global agribusiness vertical integration chain. It was the Goldberg-Davis Harvard Business School vertical integration model in spades.

By 2004, two agribusiness giants—Monsanto and DuPont's Pioneer Hi-Bred—controlled the majority of the world's private seed companies. The major GMO agribusiness companies had pursued a three-phase strategy. Initially, they either bought out or merged with most major seed companies in order to gain control over seed germplasm. Then, they took out a multitude of patents on genetic engineering techniques as well as on genetically engineered seed varieties. Finally, they required that any farmer buying their seed must first sign an agreement prohibiting the farmer from saving the seed, thus forcing them to repurchase new seed every year. In the case of Monsanto, it allowed a single company, unhampered by US Government anti-trust restrictions, to obtain unprecedented control of sale and use of crop seeds in the United States.[44]

Cleverly, the GMO seed had been marketed and developed to be resistant to that company's special herbicide. Monsanto GMO "Roundup Ready" soybeans had been genetically modified explicitly to be resistant to Monsanto's specially patented glyphosate, marketed under the brand name Roundup. They were "ready" for Roundup. That insured that farmers contracting to buy Monsanto GMO seeds must also buy Monsanto herbicide. The Roundup herbicide was so developed that it could not be used on non-GMO soybean plants. The GMO seeds were, in effect, tailor-made to fit Monsanto's existing glyphosate herbicide.

Whether such a wide proliferation of genetically modified organisms in the food chain was safe or desirable was of no concern to the agribusiness chemical and seed giants. Phil Angell, Monsanto's spokesman, put it bluntly: "Monsanto should not have to vouchsafe the safety of biotech food. Our interest is in selling as much of it as possible. Answering its safety is the FDA's job."[45]

He was well aware that the US Food and Drug Authority, on Monsanto demands, had long since abandoned any pretence of independently monitoring GMO seed safety. The Government had agreed to let the GMO companies "self-police" the industry, meaning Angell was describing a perfect circle of lies and public deception, circumscribing the incestuous relationship which had been created between the private agribusiness GMO giants and the US Government.

Lies, Damn Lies and Monsanto Lies…

The Rockefeller Foundation had carefully prepared the media marketing and propaganda case for the proliferation of genetically engineered crops. One of its main arguments was to claim that global population growth in the coming decades, in the face of gradual exhaustion of the world's best soils from over-cultivation, required a dramatic new approach to feeding the planet.

Rockefeller Foundation President, Gordon Conway, issued a public call for a second Green Revolution which he dubbed the "Gene Revolution". He insisted GMO crops were needed "to enhance food production over the next 30 years … to keep up with population increase," estimating that the world would have "an extra 2 billion mouths to feed by the year 2020." Conway further argued that GMO crops would solve the problem of how to increase crop yields on limited land, and would "avoid the problems of pesticides and the overuse of fertilizers."[46]

This carefully formulated pitch for GMO crops was picked up by the UN Food and Agriculture Organization, World Bank, IMF and the leading advocates of genetically modified seeds, especially the seed conglomerates themselves, to justify their cause. If you opposed the spread of GMO seeds, you *de facto* supported genocide against the world's poor. At least that was the not-so-subtle message of the GMO lobby.

Whether GMO crops promised large improvement in crop yield per hectare planted was also greatly disputed. Despite the most concerted effort by the GMO agribusiness companies and their financially captive university researchers, evidence began to leak into the press suggesting that those very GMO crop yields were also not what it had been cracked up to be.

A November 2004 report from the Network of Concerned Farmers in Australia concluded, in the case of genetically modified planting of canola, that

> there is no evidence that GMO canola crops yield more, but there is evidence they yield less. Although Monsanto claim a 40% yield increase with Roundup Ready canola, their best on their website for Australian trials reveal yields are 17% less than our national average. Bayer CropScience trial yields are also not comparing well against non-GMO varieties.[47]

The Soil Association in the UK issued a report in 2002 entitled "The Seeds of Doubt," based on extensive research with USA farmers who had used genetically modified crops. The report, one of the few independent assessments available, concluded that rather than boosting farmers' crop yields, "GMO soybeans and maize have worsened the situation."[48]

Based on six years of GMO growing experience, the study showed that there was real cause for alarm at the increasing farmer dependency on genetic crops. The study reported the analysis by Iowa University economist, Michael Duffy, who found that when all production factors were taken into account, "herbicide tolerant GMO soybeans lose more money per acre than non-GMO soybeans."[49]

In Argentina and Brazil, studies confirmed the emergence of glyphosate-resistant "superweeds" which were impervious to normal doses of Monsanto's Roundup glyphosate herbicide. In order to combat the damaging weeds threatening Monsanto's Roundup Ready GMO soybean fields, other supplemental herbicides had to be used. In one case in southern Brazil, where Argentine GMO seeds had been illegally smuggled in, a weed had developed which could not be killed with any dosage of glyphosate named in Brazil corda-de-viola. Only by adding DuPont's Classic herbicide would the weed die. The phenomenon became so common in fragile GMO soybean fields that a new growth segment for DuPont and the other herbicide makers became devising, patenting and producing such chemical add-ons to glyphosate. The GMO industry claims to dramatically lower herbicide requirements had been proven false.[50]

The results for genetically modified Bt corn planted in the United States were hardly better. Dr. Charles Benbrook of the Northwest Science and Environment Policy Center in Idaho, using USDA Government data in a detailed analysis of the economics of Bt corn, found that "from 1996-2001, American farmers paid at least $659 million in price premiums to plant Bt corn, while boosting their harvest by only 276 million bushels—worth $567 million in economic gain. The bottom line for farmers is a net loss of $92 million—about $1.31 per acre—from growing Bt maize (corn)."[51]

Another major drain on farmers' income, the study concluded, was the very high fees farmers had to pay to Monsanto, DuPont and the other GMO seed companies for their seeds. A significant cost was the "technology fee" charged by the seed conglomerates ostensibly to reimburse their high research and development costs.

Seeds typically accounted for 10% of normal corn production costs. GMO seeds were significantly more expensive because of the added technology fee. The study concluded that with the technology fee, "GMO seeds cost 25-40 percent more than non-GMO seeds. For Bt maize, for example, the fees are typically $8- $10/acre, about 30-35 percent higher than non-GMO varieties, though they can be up to $30/acre. Roundup Ready soybeans can have a technology fee of about $6/acre."[52] In addition, the contract prohibited farmers, at the risk of severe penalty, from reusing a portion of their seeds in the following year's planting.

Monsanto and the biotech seed giants argued that higher yields more than compensated for their added cost. Higher yields were supposedly a prime benefit of planting the GMO seeds. However, the "Seeds of Doubt" research study concluded that Monsanto Roundup Ready soybeans and Roundup Ready rapeseed produced on average lower yields than non-GMO varieties, and although genetically engineered Bt corn produced a small yield increase overall, it was not enough over the whole period to cover the higher production costs.[53]

Further contradicting the claims that GMO crops required significantly less chemical fertilizer—an argument used to win over ecological opponents—the study in fact found that Roundup Ready soybeans, corn, and rapeseed had "mostly resulted in an increase in agrochemical use," meaning more tons of pesticide and herbicide per acre than with ordinary varieties of the same crops.[54]

The study concluded that, "while there are some farmers growing GMO crops who have been able to cut their production costs or increase yields with GMO crops, it appears that, for most producers, any savings have been more than offset by the technology fees and lower market prices, as well as the lower yields and higher agrochemical use of certain GMO crops."[55]

Numerous other studies confirmed that GMO crops required not less but typically more chemical herbicides and pesticides after one or two seasons than non-GMO crops. Even the US Department of Agriculture admitted the advertised claims of GMO did not bear relation to reality. "The application of biotechnology at present is most likely ... not to increase maximum yields. More fundamental scientific breakthroughs are necessary if yields are to increase."[56]

Dr. Charles Benbrook's study, based on USDA official data, revealed that far from using less pesticides, "the planting of 550 million acres of genetically engineered corn, soybeans and cotton in the United States since 1996 has increased pesticide use by about 50 million pounds."[57]

The main reason cited for the increase was "substantial increases" in herbicide use on "herbicide tolerant"—i.e., genetically modified crops, especially soybeans, similar to the results confirmed across GMO soybean fields in Brazil and Argentina. There was a significant increase in herbicide use on GMO crops compared to acres planted with conventional plant varieties. "Herbicide tolerant" plants were genetically modified to ensure that those who grew the crops had no other option but to also use the herbicides of the same companies.

Farmers across the United States, where GMO crops had been planted for a number of years, discovered that, unexpectedly, herbicide-tolerant weeds had emerged requiring added use of other herbicides in addition to the GMO-specific brands such as Monsanto's Roundup Ready.[58] In the case of GMO corn, the weed plague had necessitated the use of the chemical herbicide atrazine, one of the most toxic herbicides that exist, as a supplement to control weeds. Many independent crop scientists and farmers predicted the imminent danger of the creation of super-weeds and Bt-resistant pests which could threaten entire harvests.

Increasingly, it appeared that the case in favour of widespread commercial use of genetically engineered seeds for agriculture had been based on a citadel of scientific fraud and corporate lies.

GMO Soybeans and Infant Death?

From Russian science came another test, the results of which were attacked and belittled by the marvellous propaganda machine of the GMO agribusiness lobby.

In January 2006, a respected London newspaper, *The Independent* carried a story titled, "Unborn Babies Could be Harmed by GMOs."[59] The article reported on research results from scientist Dr. Irina Ermakova of the Institute of Higher Nervous Activity and Neurophysiology of the Russian Academy of Sciences.

Her study found that more than half of the offspring of rats fed on genetically modified soybean diet died in the first three weeks of life—six times as many as those born to mothers with normal diets.

Dr. Ermakova added flour from Monsanto's GMO soybean to the food of female rats, starting two weeks before they conceived, continuing through pregnancy, birth and nursing. Others were given non-GM soybeans, and a third group was given no soybean at all.

The Russian scientist was alarmed to find that 36 percent of the young of rats fed the diet of modified soybeans were severely underweight, compared to 6 percent of the offspring of the other groups. More alarmingly, a staggering 55.6 percent of those born to mothers on the GMO diet died within three weeks of birth, compared to 9 percent of the offspring of those fed normal soybeans, and 6.8 percent of the young of those given no soybeans at all. "The morphology and biochemical structures of rats are very similar to those of humans, and this makes the results very disturbing," said Dr Ermakova. "They point to a risk for mothers and their babies."[60]

Monsanto and other GMO firms attacked the credibility of Dr. Ermakova while curiously avoiding the obvious call to repeat the simple test in other labs to confirm or refute. Monsanto's Public Relations department merely restated their mantra. Tony Coombes, director of corporate affairs for Monsanto UK, told the press, "The overwhelming weight of evidence from published, peer-reviewed, independently conducted scientific studies demonstrates that Roundup Ready soy can be safely consumed by rats, as well as all other animal species studied."

The Russian results were potentially so serious that the American Academy of Environmental Medicine asked the US National Institute of Health to sponsor an immediate, independent follow-up.[61]

Africa's Fake "Wonder Potato"

In one of its more widely publicized acts, Monsanto donated a genetically engineered virus-resistant sweet potato in Africa to the Kenya Agricultural Research Institute (KARI), an institute supported financially by, among others, the World Bank and Monsanto.

KARI's Dr. Florence Wambugu was deployed by Monsanto and USAID to give talks around the world, where she proclaimed that the GMO sweet potato of Monsanto had solved hunger in Africa. The GMO sweet potato had been developed by Wambugu while at Monsanto in St. Louis in a project backed by the USAID, ISAAA and the the World Bank. Wambugu claimed it would raise yields from four tons to ten tons per hectare.[62] In 2001, USAID backed the project through a high profile promotion to spread GMO crops on a skeptical African population. *Forbes*, an American financial magazine which referred to itself as, "The Capitalist Tool," proclaimed Wambugu as one of 15 people from around the world who would "re-invent the future".[63]

The only problem was that the re-inventing project was a catastrophic failure. The GMO sweet potatoes proved susceptible to viral attacks. Their yields were proven less than that of normal indigenous sweet potatoes, not 250% greater as predicted by Wambugu.[64] KARI and its corporate backers tried to maintain the fraud, but Dr. Aaron deGrassi of the Sussex University Institute of Development Studies exposed the statistical gimmicks being used by Wambugu and Monsanto to claim their gains.

DeGrassi stated that, "Accounts of the transgenic (GMO-w.e.) sweet potato have used low figures on average yields in Kenya to paint a picture of stagnation". An early article stated 6 tons per hectare-without mentioning the data source-which was then reproduced in subsequent analyses. However, deGrassi noted, "FAO statistics indicate 9.7 tons, and official statistics report 10.4 tons."[65]

The World Bank and Monsanto ignored the critical findings and continued financing Wambugu's research for more than 12 years.

As the late American humorist and social critic, Mark Twain, might have said of the situation: "There are three kinds of lies: Lies, Damn Lies and Monsanto lies...."

In the heady environment of a US biotech stock market euphoria at the end of the 1990's, and the falling barriers to GMO proliferation, Monsanto, Syngenta and the major seed giants nearly ran off the rails with their project to take over the seed supply of the world. It required an extraordinary intervention in 1999 by their patron saint, the Rockefeller Foundation, to rescue the over-eager agribusiness giants from their own methods.

Notes

1. Letter from Daniel G. Amstutz, Undersecretary for International Affairs and Commodity Programs, U.S. Department of Agriculture, published in *Choices*, Fourth Quarter, 1986, p. 38.

2. *Who's Who in Corporate Agribusiness*, http://www.electricarrow.com/CARP/tiller/archives/backlog.htm, 3 April 1997.

3. Eugene W. Plawiuk, "Background on Cargill Inc., the Transnational Agribusiness Giant", *Corporate Watch: GE Briefings*, http://www.archive.corporatewatch.org, November 1998.

4. *Ibid.*

5. Lori Wallach and Michelle Sforza, *The WTO: Five Years of Reasons to Resist Corporate Globalization*, Seven Stories Press, New York, 1999, p. 45.

6. Edward A. Evans, "Understanding the WTO Sanitary and Phytosanitary Agreement", *EDIS document FE492*, Department of Food and Resource Economics, Florida Cooperative Extension Service, UF/IFAS, University of Florida, Gainesville, FL., August 2004, http://edis.ifas.ufl.edu. Evans notes: "The United States feared that, with a reduction in the use and levels of these support measures, some importing countries might turn to technical trade barriers (notably SPS measures) as a means of allowing them to continue providing support to their farming community. Consequently, the intent of the Agreement was to ensure that when SPS measures were applied, they were used only to the extent necessary to ensure food safety and animal and plant health, and not to unduly restrict market access for other countries." Who would determine "extent necessary" were the tribunals of the WTO.

7. World Development Movement, *GMOs and the WTO: Overruling the Right to Say No*, http://www.wdm.org.uk, London, November 1999. See also, Edward A. Evans, *op.cit.*

8. World Development Movement, *op.cit.*

9. Anup Shah, *Food Patents—Stealing Indigenous Knowledge?*, http://www.global-issues.org, 26 September 2002.

10. GRAIN, *Genetech Preys on the Paddy Field*, http://www.grain.org/seedling/?id=33, June 1998. See also *Thai Jasmine Rice and the Threat of the US Biotech Industry*, News Release, http://www.biotech-info.net/wescott_thai_rice.pdf.

11. GRAIN, *op. cit.* The report notes, "Although it is IRRI's policy not to patent either the germplasm collected from farmers' fields or the products of its conventional breeding work, it cannot prevent those who access their collections from doing so. This laissez faire policy has translated into the appropriation of part of farmers' rice germplasm by the private sector. For example, a US breeding company,

Farms of Texas, made some minor modifications on IRRI's IR8 and patented it for exclusive sale in the United States. In early 1998, that company, now named RiceTec, caused public outrage by obtaining a patent on Indian and Pakistani Basmati rice."

12. United States Supreme Court, *J.E.M. Ag Supply V. Pioneer Hi-Bred*, 122 S.Ct. 593, 2001. Also for background, "CAFC Decision in Pioneer Hi-Bred International, Inc. vs. J.E.M. Ag Supply, Inc. et al.", *Biotechnology Law Report*, April 2000, pp. 281-289. The issue which the Supreme Court decised in favor of Pioneer Hi-Bred's claim to patent rights for genetically manipulated plants was accepting a ruling of the US Court of Appeals, which had upheld a lower court ruling that patents on Pioneer genetically engineered corn seed were valid. The court concluded that title 35, section 101 of the United States Code (part of the Patent Act) "includes seeds and seed grown plants." This section states that "[w]ho[m]ever invents or discovers any new and useful process, machine, manufacture, or composition of matter, or any new and useful improvement thereof," can get a patent to protect their work. The Supreme Court ruled in favor of Pioneer, in effect upholding the right of a seed company to achieve iron clad protection for its bioengineered products by allowing general utility patents in addition to a PVPA (Plant Variety Protection Act) certificate." Also, Edmund J. Sease presents an excellent legal review of the process in Edmund J, Sease, "History and Trends in Agricultural Biotechnology Patent Law from a Litigator's Perspective", *Seeds of Change Symposium Banquet*, http://www.ipagcon.uiuc.edu, University of Illinois, Chicago, 9 April 2004.

13. Robert B. Shapiro, cited in Richard A. Melcher et al., "Fields of Genes", *Business Week*, 12 April 1999.

14. John Vidal, "Monsanto Dumped Toxic Waste in UK", *The Guardian*, 12 February 2007.

15. Meryl Nass, *op. cit.*

16. Press Release, *DuPont and Pioneer Hi-Bred International, Inc., Merger Completed*, Delaware, 1 October 1999.

17. Dow Chemical Company, *Quarterly Report*, in http://www.dow.com/financial/reports/07q1earn.htm.

18. Michigan Department of Environmental Quality, "Soil Movement Advisory", *Environmental Assessment Initiative*, Information Bulletin #3, June 2003.

19. *Ibid.*

20. Testimonials from Dow Midland employees about effects of dioxin are available on http://www.studentsforbhopal.org. On documentation of Dow Corning's defective silicone breast implants, for which the company had to pay $3.5 billion

in damages to thousands of victims, see *Implant Veterans of Toxic Exposure*, yukonmom47.tripod.com/index.html.

21. Quoted in Arundhati Roy, "The Loneliness of Noam Chomsky", *The Hindu*, 24 August 2003. A US Vietnam veteran is attributed with this perverse but descriptive quote about the development of Napalm: "We sure are pleased with those backroom boys at Dow. The original product wasn't so hot—if the gooks [Vietnamese] were quick they could scrape it off. So the boys started adding polystyrene—now it sticks like shit to a blanket. But then if the gooks jumped under water it stopped burning, so they started adding Willie Peter (white phosphorus) so's to make it burn better. It'll burn under water now. And just one drop is enough; it'll keep on burning right down to the bone so they die anyway from phosphorus poisoning." See also "Protesting Napalm", *Time*, 5 January 1968. During the rising antiwar protests which cost Lyndon Johnson his Presidency, Dow Chemical became the symbol of the brutality of the American war machine. The company refused to cease production of napalm, despite the fact it represented 0.5% of gross revenues.

22. *Dow Agrosciences Llc, Plaintiff-Appellee Versus Dennis Bates* et al., United States Court of Appeals, Fifth Circuit, 11 June 2003.

23. Der Spiegel, 8 February 2004.

24. Gottfried Glöckner as interviewed by Klaus Faissner, "Der Genmais und das grosse Rindersterben", *Neue Bauernkoordination Schweiz*, http://www.nbks.ch/abstim/gentechmais_rindersterben.html#.

25. Gottfried Glöckner, in a letter to the author, *Gentechnik Mais*, 3 February 2007.

26. Klaus Faissner, *Gentechnik oder Bauern?*, http://www.arge-ja.at/gentechnik_landwirtschaft_faissner.html.

27. ETC Group Communique, *Syngenta—The Genome Giant?*, http://www.etc-group.org., January/February 2005.

28. *Investigate*, New Zealand, January/February 2001.

29. Elmo R. Zumwalt, quoted in Meryl Nass, *Monsanto's Agent Orange: The Persistent Ghost from the Vietnam War*, http://www.organicconsumers.org/monsanto/agentorange032102.cfm#what.

30. Cate Jenkins, "Criminal Investigation of Monsanto Corporation—Cover-up of Dioxin Contamination in Products—Falsification of Dioxin Health Studies", *USEPA Regulatory Development Branch*, November 1990. See also "The Legacy of Agent Orange", *BBC News, World Edition*, 29 April 2005, transcript in home.clara.net/heureka/gaia/orange.htm.

31. BBC News, *op. cit.*

32. *Ibid.*

33. Cate Jenkins, quoted in "Fields of Genes: The Battle over Biotech Foods", *This Morning*, CBC Radio, 3-7 May 1999. "Fields of Genes part 4", 6 May 1999, http://www.nyenvirolaw.org/PDF/CBC-05-06-1999-FieldOfGenes-TheBattleOverBiotechFoods.PDF.

34. The Monsanto Corporation has been accused of not listening to the groups that would be responsible for marketing their Roundup Ready GMO soybean, as they shipped tons of these beans to soy processors in Europe; in *Nature Biotechnology* 14 (1996), 1627; Nature 384 (1996), 203, 301; NS (7 Dec 1996), 5: "A trade war nearly began between USA and Europe. Hans Kroner, secretary-general of Eurocommerce, representing retailers in 20 European countries, recently called for Roundup Ready to be segregated from other beans. Earlier in 1996, European retail and wholesale groups had asked for separate streams for the Roundup Ready. Retailers in France, Denmark, the Netherlands, and the United Kingdom wanted segregation so that they could label the products appropriately. German, Austrian, Finnish, and Swedish retailers wanted a separate stream so that they could exclude genetically manipulated food either 'for the foreseeable future' or 'until consumers are happy'." Cited in *Food Safety Including GM Foods*, eubios.info/NBB/NBBFS.htm.

35. Clive James, "Global Status of Commercialized Biotech/GM Crops: 2004", *IS AAA*, No. 32, 2004.

36. *Ibid.*

37. Sally Falon and Mary Enig, "The Great Con-ola", *Nexus Magazine*, http://www.nexusmagazine.com, August-September 2002. The authors conclude: "canola oil is definitely not healthy for the cardiovascular system. Like grapeseed oil, its predecessor, canola oil is associated with fibrotic lesions of the heart. It also causes vitamin E deficiency, undesirable changes in the blood platelets, and shortened life-span in stroke-prone rats when it was the only oil in the animals' diet. Furthermore, it seems to retard growth, which is why the FDA does not allow the use of canola oil in infant formula." http://www.nexusmagazine.com/articles/canola.html

38. The Pew Charitable Trusts, *Genetically Modified Crops in the United States*, pewagbiotech.org, August 2004.

39. Mae-Wan Ho and Lim Li Ching, "The Case for A GM-Free Sustainable World", *Independent Science Panel*, London, 15 June 2003, p. 23, in http://www.food-first.org/progs/global/ge/isp/ispreport.pdf.

40. *Ibid.*

41. *Ibid.*

42. *Ibid.*

43. Andrew Kimbrell, "Monsanto vs. US Farmers", *The Center for Food Safety*, http://www.centerforfoodsafety.org/Monsantovsusfarmersreport.cfm., Washington DC, 2005, pp. 19-21.

44. *Ibid.*, pp. 7-11.

45. Phil Angell, quoted in "Playing God in the Garden", *New York Times Magazine*, 25 October 1998. That was a direct contradiction of at least officially published US government policy: "Ultimately, it is the food producer who is responsible for assuring safety"—Food and Drug Administration, *Statement of Policy: Foods Derived from New Plant Varieties* (GMO Policy), Federal Register, Vol. 57, No. 104, 1992, p. 229.

46. Gordon Conway, *The Rockefeller Foundation and Plant Biology*, speech, http://www.biotech-info.net/gordon_conway.html, 24 June 1999.

47. Network of Concerned Farmers, *Will GM Crops Yield More in Australia?*, http://www.non-gm-farmers.com, 28 November 2004.

48. Gundula Meziani, and Hugh Warwick, "The Seeds of Doubt", *The Soil Association*, http://www.soilassociation.org, 17 September 2002. A very comprehensive scientific independent review of the claims and counter-claims of GMO is the study of Mae-Wan Ho and Lim Li Ching, *op. cit.*

49. Gundula Meziani, *op. cit.*

50. Antonio Andrioli, Universidade Regional do Noroeste do Estado do Rio Grande do Sul, Brazil, *Private e-mail Correspondence*, 27 January 2007, provided on request by Friedel Kappes.

51. Charles M. Benbrook, "Biotech Crops Won't Feed Africa's Hungry", *The New York Times*, 11 July 2003.

52. *Ibid.*

53. Gundula Meziani, et al., *op. cit.*, pp. 19-20.

54. *Ibid.* p. 19.

55. *Ibid.* p. 20.

56. USDA, *Agriculture Information Bulletin*, 2001.

57. Gundula Meziani, et al., *op. cit.*, p. 19.

58. *Ibid.*, pp. 19-20.

59. Geoffrey Bean, "Unborn Babies Could be Harmed by GMOs", *The Independent*, 8 January 2006.

60. *Ibid.*

61. *Ibid.*

62. Gundula Meziani, *op. cit.*

63. Lynn J. Cook, "Millions Served", *Forbes*, 23 December 2002.

64. *Ibid.* See also, GM Watch, *Wambugu Wambuzling Again: Says GM Sweet Potato a Resounding Success?*, http://www.mindfully.org/GE/2004/Wambugu-Wambuzling-Again17mar04.htm, 17 March 2004. Also, "Monsanto's Showcase Project in Africa Fails", *New Scientist*, 7 February 2004. The article notes, "Three years of field trials have shown that GM sweet potatoes modified to resist a virus were no less vulnerable than ordinary varieties, and sometimes their yield was lower, according to the Kenya Agricultural Research Institute. Embarrassingly, in Uganda conventional breeding has produced a high-yielding variety more quickly and more cheaply. The GM project has cost Monsanto, the World Bank and the US government an estimated $6 million over the past decade. It has been held up worldwide as an example of how GM crops will help revolutionize farming in Africa. One of the project members, Kenyan biotechnologist Florence Wambugu (see *New Scientist*, 27 May 2000, p. 40), toured the world promoting the work."

65. GM Watch, *op. cit.*

PART V
Population Control

CHAPTER 12
Terminators, Traitors, Spermicidal Corn

"Two Steps Forward, Then One Step Backward…"

By the end of the 1980's, backed by the new clout of the WTO and the full support from the White House, the genetic seed giants began to get visibly intoxicated by the possibilities of taking over the world's food supply. They all were working feverishly on a new technology which would allow them to sell seed that would not reproduce. The seed companies named their innovation GURTs, short for Genetic Use Restriction Technologies.

The process was soon known as "Terminator" seeds, a reference to Arnold Schwarzenegger's crude and death-ridden Hollywood films. As one GMO Terminator backer put it, it was developed to "protect corporations from unscrupulous farmers" (*sic*) who might try to re-use patented seed without paying. No matter that the vast majority of the world's farmers were too poor to afford the Monsanto GMO license and other seed fees, and had re-used seed for thousands of years before.

In 1998, Delta & Pine Land Seed Company, a US bio-tech company in Scott, Mississippi, was the largest owner of commercial cotton seeds. With financial backing from the US Department of

Agriculture, it had won a joint patent together with the US Government, for its GURT, or Terminator, technology. Their joint patent, US patent number 5,723,765 titled "Control of Plant Gene Expression," allowed its owners and licensees to create sterile seed by selectively programming a plant's DNA to kill its own embryos. The patent applied to plants and seeds of all species.[1]

If farmers tried to save the seeds at harvest for future crops, the seeds produced by these plants would not grow. Peas, tomatoes, peppers, wheat, rice or corn would essentially become seed cemeteries. As one critic put it, "In one broad, brazen stroke of his hand, man will have irretrievably broken the plant-to-seed-to plant-to-seed cycle, the cycle that supports most life on the planet. No seed, no food … unless you buy more seed."[2]

One year later Monsanto announced it was buying Delta & Pine Land. They had their eyes firmly on getting the Terminator patent. They knew it was applicable not only to cotton seeds but to all seeds.

Terminator looked like the answer to the agribusiness dream of controlling world food production. No longer would they need to hire expensive detectives to spy on whether farmers were re-using Monsanto seeds.

Terminator corn, soybeans, or cotton seeds had been genetically modified to "commit suicide" after one harvest season. The inbuilt gene produced a toxin just before the seed ripened, whereby in every seed the plant embryo would self-destruct. The Terminator seeds would automatically prevent farmers from saving and re-using the seed for the next harvest. The technology was a beautiful means of enforcing Monsanto or other GMO patent rights and fees, especially in developing economies where patent rights were little respected.

A second, closely related technology which held priority R&D funding by the gene multinationals in the late 1990's was T-GURT seeds, the second generation of Terminator. T-Gurts, or Trait Genetic Use Restriction Technologies, were nicknamed "Traitor," a reference to the plant trait features of the genetic technology used. It was also a word which had a double meaning not lost on its critics.

Traitor technologies relied on controlling not only the plant's fertility, but also its genetic characteristics. In its US patent application, Delta & Pine Land and the USDA stated the method with "an inducible gene promoter that is responsive to an exogenous chemical inducer," called a "gene switch." This promoter can be linked to a gene and introduced into a plant. The gene can be selectively expressed (i.e. activated) by application of the chemical inducer to activate the promoter directly.

The official patent application continued. Growth of the plant can be controlled by the application or withholding of a chemical inducer. While the inducer is present, the repressor is expressed, the promoter attached to the disrupter gene is repressed, the disrupter protein is not expressed, thereby allowing the plant to grow normally. If the chemical inducer is withheld, the gene switch is turned off, the repressible promoter is not repressed, so the disrupter protein is expressed and plant development is disrupted.[3]

A GMO crop of rice or corn would only be resistant to certain plagues or pests after use of a specific chemical compound, which would only be available from Monsanto, Syngenta, or other owners of patent rights to the specific Traitor seeds. Farmers trying to buy seed from the "illegal" seed market would not be able to get the special chemical compound needed to "turn on" the plant's resistance gene.

Traitor technology offered a unique chance to open an entire new captive market for Monsanto and the others to sell their agrichemicals. Furthermore, Traitor was cheaper to produce than the complicated Terminator seeds. Not widely publicized, the fact about Traitor technologies was that with them it was also possible to develop GMO plants that needed to be "turned on" in order to grow or become fertile.

One study noted that 11 new patents were held by the newly-formed Syngenta. These patents allowed "genetic modification of staple crops which will produce disease prone plants (unless treated with chemicals); control the fertility of crops; control when plants flower; control when crops sprout; control how crops age."[4]

By the year 2000 Syngenta had the single largest interest in GURTs of all the global GMO companies. Monsanto was determined to change that, however.[5]

Under the Terminator joint agreement between the USDA and Delta & Pine Land, D&PL had exclusive licensing rights, while the USDA would earn about 5 percent of the net sales of any commercial product using the technology. The USDA and Pine Land Co. also applied for patents in some 78 other countries. The official backing of the US Government gave the patent application huge leverage that a small private company would lack abroad. Delta & Pine Land said in its press release that the technology had "the prospect of opening significant worldwide seed markets to the sale of transgenic technology for crops in which seed currently is saved and used in subsequent plantings."[6]

In practice, farmers purchased elite seeds that provided only one harvest; the seeds from this harvest were sterile, absent, or non-elite and the farmer must buy either seed or trait-maintenance chemical compound from the company.[7]

The US Government defended its patent on GURTs, which they named TPS for the benign-sounding "Technology Protection System":

> Because of this seed-saving practice, companies are often reluctant to make research investments in many crops; they cannot recoup their multi-year investment in developing improved varieties through sales in one year. TPS would protect investments made in breeding or genetically engineering these crops. It would do this by reducing potential sales losses from unauthorized reproduction and sale of seed.[8]

At the time, in a revealing but little-noticed statement, Delta & Pine Land admitted that the initial reason they developed Terminator technology was to market it to rice and wheat farmers in countries such as India, Pakistan and China.

The implications of Terminator and Traitor technology in the hands of the GMO agribusiness giants were difficult to grasp. For the first time in history, it would allow three or four private multinational seed companies to dictate terms to world farmers for their seed. There are several major crops which usually are not grown from hybrid seeds. These include wheat, rice, soybeans, and cotton. Farmers often save the seeds from these crops, and may not need to go back to the seed company for several years—or longer, in some parts of the world—to purchase a new variety.[9]

In the hands of one or more governments intent on using food as a weapon, Terminator was a tool of biological warfare almost "too good to believe." In their US patent applications, the companies stated, "seed savers number an estimated 1.4 billion farmers worldwide—100 million in Latin America, 300 million in Africa, and 1 billion in Asia—and are responsible for growing between 15 and 20 percent of the world's food supply."[10]

The Guardian Angel Saves the GMO Project

An ensuing public uproar over the prospect of major private seed multinationals controlling seeds through Terminator technology threatened the very future of the entire Gene Revolution. Ministers were delivering Sunday sermons on the moral implications of Terminator; farmers were organizing protests; governments were holding public hearings on the new development in gene technology. Across the European Union, citizens were in open opposition to GMO because of the Terminator threat and its implications for food security, and because of the fact that the US and other patent offices had decided to grant exclusive patents to Monsanto and Syngenta for several different varieties of Terminator.

The widespread and growing protest against the obvious potential for misuse of Terminator suicide seeds took on a new character in May 1998. Monsanto, which had already gotten one patent on Terminator gene technology six months earlier, announced it would buy Delta & Pine Land. The move would make Monsanto the unquestioned leader in genetic Terminator technology.

News of the planned takeover became a public relations disaster for Monsanto. Newspaper headlines around the world portrayed it as exactly what it was—an attempt by a private corporation to control the seed supply of world farmers.

The growing opposition to genetically modified foods, fed by the negative publicity given to the Terminator seed, led to a dramatic intervention by the guardian angel of the GMO global project. In September 1999, Gordon Conway, the Rockefeller Foundation President, took the highly unusual step of asking to personally address the Board of Directors of Monsanto. He made clear to them that what was at stake was to demand Monsanto not to persist in developing and commercializing Terminator seed technologies.[11]

Monsanto listened carefully to Conway. On October 4, 1999, Monsanto CEO, Robert B. Shapiro, held a press conference where he announced that the company had decided to stop the process of commercializing the Terminator technology. Shapiro repeated his position in an Open Letter that month to Rockefeller Foundation President Conway, where he said, "We are making a public commitment not to commercialize sterile seed technologies, such as the one dubbed "Terminator." We are doing this based on input from you and a wide range of other experts and stakeholders." The world's press covered it as a major victory for the side of reason and social justice. In reality, it was a shrewd tactical deception, worked out together with Rockefeller Foundation's Conway.

For those who bothered to read the fine print, Monsanto had in fact given up nothing. Monsanto's Shapiro did not back off or reject the chance to develop Terminator in the future. Only for an undefined time would there be a moratorium on "commercialization." The commercial stage of Terminator at that point was believed at least several more years away, earliest perhaps in 2007, so little would be lost for Monsanto and much would be won in terms of public relations.

Shapiro made clear in his public statement that he was not about to give up such a weapon over seed supply without a fight. He declared that, "Monsanto holds patents on technological approaches to gene protection that do not render seeds sterile and has studied one that would inactivate the specific gene responsible for the value-added biotech trait."[12] He was referring to Traitor technologies. Shapiro added that, "We are not currently investing resources to develop these technologies."[13]

"But," he stressed, *"we do not rule out their future development and use for gene protection or their possible agronomic benefits"* (emphasis added). Shortly after that statement, Monsanto announced that it had called off plans to take over Delta & Pine Land as well. All appeared to signal the death of Terminator.[14]

Syngenta announced at the same time that it was also declaring a moratorium on the commercialization of Terminator, adding that it would, however, continue with its Traitor developments. The heat was off the Terminator controversy; the deception had apparently worked, as press headlines about Terminator began to disappear from view.

Notably, while Rockefeller's Conway and Monsanto Corporation were making headlines with their declarations on Terminator suspension, the US Department of Agriculture, the partner in Terminator with Delta & Pine Land, made no such commitment. This was indeed curious, as it would have been easy and uncomplicated for the USDA to follow the gene giants by declaring its own moratorium. The press paid no attention to this. Monsanto's news was the headline story.

In a June 1998 interview, USDA spokesman Willard Phelps had declared the US Government policy on Terminator seeds. He explained that the USDA wanted the technology to be "widely licensed and made expeditiously available to many seed companies." He added that the Government's aim was "to increase the value of proprietary seed owned by US seed companies and to open up new markets in Second and Third World countries." The USDA was open about its reasons. It wanted to get Terminator seeds into the developing world, where the Rockefeller Foundation had put eventual proliferation of genetically engineered crops at the heart of its GMO strategy from the beginnings of its rice genome project in 1984.[15]

The Terminator technology was being supported at the highest levels of the US government to target agriculture in the Second and Third World. It would make it "safe" for Monsanto, DuPont and the other seed giants to market their GMO seeds in targeted developing countries. The USDA microbiologist primarily responsible for developing Terminator with D&PL, Melvin Oliver, openly admitted: "My main interest is the protection of American technology. Our mission is to protect US agriculture, and to make us competitive in the face of foreign competition. Without this, there is no way of protecting the technology [patented seed]."[16]

Together with Delta & Pine Land, the USDA had applied for Terminator patents in 78 countries. The USDA admitted openly, perhaps carelessly so, that the target for Terminator seeds were the populations and farmers of the developing world, precisely the Rockefeller Foundation's long-standing goal of promoting GMO.

The coherence between the 1974 Henry Kissinger NSSM 200 population control policies in the developing world, the Rockefeller Foundation's support for introduction of gene technologies in targeted developing countries, and the development of a technology which would allow the private multinationals owning the patents on vital staple seed varieties, was also beginning to dawn on a broader thinking public. The development by Monsanto was increasingly being seen by the world as a kind of Trojan Horse for Western GMO seed giants to get control over Third World food supplies in areas with weak or non-existent patent laws.

The Rockefeller-Monsanto public moratorium announcement in October 1999 was a calculated ploy to direct attention elsewhere, while the seed companies continued their perfection of Terminator, Traitor and related technologies.

Meanwhile, as the Rockefeller Foundation understood it, the urgent priority for the time was to spread GMO seeds worldwide in order, first, to capture huge markets and to make the use of patented GMO seeds irreversible. In some cases, companies like Monsanto were accused by local farmers of illegally smuggling GMO seeds into regions like Brazil or Poland, in order to later claim that farmers had "illegally" used their patented seed, while demanding they pay royalties.

In the case of Brazil, Monsanto was shrewd. Monsanto used the smuggling of GMO soybeans to its advantage, working with the illegal GMO soy producers to pressure the Lula da Silva government to legalise the crop. Once the GMO soy became legal in Brazil,

Monsanto moved in to put an end to the "black market". With the government offering an amnesty to farmers who registered their crops as GMO soy, Monsanto worked out an agreement with producer organisations and soybean crushers, cooperatives and exporters, to force Brazilian farmers to pay royalties.[17]

Rockefeller's Conway clearly realized that the entire strategy of achieving global control over the food supply was being jeopardized in its most fragile stages by the relentless drive of Monsanto to promote its Terminator technology. In 1999, GMO seeds had barely assumed a significant share in the US seed market. Their proliferation in developing countries, with occasional exceptions such as Argentina, was at that time minimal. The European Union had imposed a ban or moratorium on licensing GMO plants. Brazil, Mexico and many African nations had strict bans on GMO imports or cultivation. The entire Gene Revolution project of the Rockefeller Foundation and its corporate and political allies was in danger of flying off the track if Monsanto persisted with the public development of Terminator.

Were the world to wake up to what was possible with GMO seeds, it might rebel while it still could. This was the evident reasoning behind the rare event of the Rockefeller Foundation's public intervention. In order to save the entire project, Rockefeller in effect imposed a higher discipline on Monsanto, and Monsanto got the message.

Terminator developments never stopped after 1999.

While Monsanto did abandon merger talks with Delta & Pine Land in late 1999, Delta & Pine and the USDA continued with their full program to perfect Terminator and Traitor technologies. Delta Vice President, Harry Collins, declared in a press interview to his GMO trade peers in the *Agra/Industrial Biotechnology Legal Letter*, "We've continued right on with work on the Technology Protection System (TPS or Terminator). We never really slowed down. We're on target, moving ahead to commercialize it. We never really backed off."[18]

Neither did their partner, the United States Department of Agriculture, back down after 1999. In 2001, the USDA Agricultural

Research Service (ARS) website announced: "USDA has no plans to introduce TPS into any germplasm .... Our involvement has been to help develop the technology, not to assist companies to use it"—as if to say, "our hands are clean."[19]

They weren't.

The USDA went on to say it was "committed to making the [Terminator] technology as widely available as possible, so that its benefits will accrue to all segments of society .... ARS intends to do research on other applications of this unique gene control discovery .... When new applications are at the appropriate stage of development, this technology will also be transferred to the private sector for commercial application."[20] Terminator was alive and well in the hands of US Government.

In August 2001, the USDA announced that it had signed a license agreement with its partner, Delta & Pine Land, allowing D&PL to commercialize the Terminator technology for its cotton seeds. The public outcry this time was mute. The issue had fallen off the public radar screen, and days later the events of September 11, 2001 completely buried the USDA announcement. The world suddenly had other concerns.

After the Terminator furor had died down in June 2003, Monsanto had begun to repackage Terminator as an "ecological plus." Rather than stress the seed control aspect, Monsanto began to promote Terminator or GURTs as a way to control the spread of GMO seeds by wind or pollination and the contamination of non-GMO crops. In February 2004, Roger Krueger of Monsanto wrote a paper together with Harry Collins of Delta & Pine Land in the magazine of the International Seed Federation, the umbrella association for the industry. Their article dismissed all worries about the dangers of Terminator or GURT seeds as "conjecture," and declared that "GURTs have the potential to benefit farmers in all size, economic and geographical areas." This time they referred to Terminator or GURTs as "a possible technical solution" to problems of plant contamination.

"Push it Down Their Bloody Throats…"

As soon as the furor around Terminator seeds had vanished from the headlines of the world's press, the major gene seed companies, in concert with the US Government, began employing increasingly coercive tactics to force GMO seeds down the throats of the world population, especially in the developing world. Among their techniques of persuasion, the genetic seed companies employed bribery, coercion, and illegal smuggling of their GMO seeds into country after country to "spread the Gospel of GMO salvation."

In 2002, the State Department instructed all its aid agencies to act as international policemen. They were instructed by USAID, a Government agency, to immediately report to them any opposition in a recipient country, to GMO food imports. They were told to collect documentation to determine if the anti-GMO attitude of the local government was "trade or politically motivated." If they determined it was trade motivated, the US Government had recourse to the WTO or to the threat of WTO sanctions against the aid recipient country, usually a potent threat against poor countries.[21]

To help Monsanto, DuPont and the other US seed giants spread GMO seeds. The US State Department and US Department of Agriculture coordinated to give emergency famine relief aid in the form of genetically modified US surplus commodities, a practice condemned by international aid organizations, as it destroyed a country's local agricultural economy in the process of opening new markets for Monsanto and friends. The European Union issued an official protest at the US Government's "use of food aid donations used as surplus disposal measures."[22] Washington ignored the protest.

In early 2003, the Government of India refused to allow import of 1,000 tons of US soybean-corn blend on the grounds that it might contain genetically modified foods which could be hazardous to human health and which had not been adequately tested. The import, through the American food aid organizations CARE and Catholic Relief Services, had thus not been approved. USAID ignored that small fact and pressed ahead.[23]

The practice long established by international aid agencies was to buy their food supplies on the open market, if possible from farmers in the recipient country or in neighboring countries. USAID mandated that US-based food aid organizations ship only grain provided by USAID, which meant genetically modified US grain. The United States was practically the only donor country insisting on use of its own food surplus for food aid.

In October 2002, the London *Guardian* reported that the US Government offered emergency famine relief aid during a severe drought, aid worth $266 million, to six countries in southern Africa. However, it offered it only in the form of genetically modified corn from US surplus stocks, although ample conventional corn was available on the market.[24] Corn was the staple food in that region of Africa. Zambia, Malawi and Zimbabwe all refused the GMO corn, citing possible health hazards. EU and other food aid donors gave the respective countries cash to buy their food on the open market instead, the customary international practice in such situations of famine. Washington had another agenda: spread the use of GMO seeds as far and wide and as fast as possible, by whatever means necessary.

When USAID Administrator, Andrew Natsios, was questioned by the press, he snapped back, "starving people do not plant seeds. They eat them."[25] The farmers where the GMO seeds were taken, of course, planted the seed for a next harvest, unaware for the most part of what seed they had gotten. It carried no GMO label, as Monsanto or DuPont or another of the seed giants would later remind them. The UN claimed that 160,000 tons of non-GMO cereals, including corn, were available in neighbouring South Africa, Kenya and nearby states for relief aid.[26]

Referring to the USAID pressure on Zambia to accept US GMO corn as famine aid, Dr. Charles Benbrook, an agronomist and former Executive Director of the US National Academy of Sciences Board on Agriculture, replied that, "[t]here is no shortage of non-GMO foods which could be offered to Zambia, and to use the needs of Zambians to score 'political points' on behalf of biotechnology was unethical and indeed shameless."[27]

In 2001, the International Monetary Fund and World Bank, two organizations dominated by Washington, demanded that Malawi's Government sell off its state emergency food reserves in order to repay their foreign debts due in 2002. Predictably, in the midst of a severe drought, Malawi had no food to feed its starving population. USAID shipped 250,000 metric tons of surplus US GMO corn. Professor David King, Science Adviser to Britain's Prime Minister, denounced the efforts by the United States to force GMO technology into Africa, calling it a "massive human experiment." The British aid organization, ActionAid, criticized the US action, declaring, "farmers will be caught in a vicious circle, increasingly dependent on a small number of giant multinationals for patented seeds."[28]

That was precisely the plan.

George W. Bush threw the considerable weight of his office to back the campaign at a G8 European Summit in June 2003, in which he stated:

> Our partners in Europe have blocked all new bio-crops because of unfounded, unscientific fears. This has caused many African nations to avoid investing in bio-technologies for fear that their products will be shut out of European markets.[29]

Bush was upping the heat on the EU to lift its 1997 ban on GMO plants. Southern Africa had some of the richest most fertile soil in the world, abundant supplies of fresh water and a benign climate. Agribusiness companies like Monsanto and Cargill were clearly salivating at the prospects of using their industrial factory farming and GMO plant cultivation. Only a few tens of millions of poor African citizens stood in the way.

However, Africa was not the only target for the worldwide proliferation of GMO seeds in the early months of the new millennium. Monsanto, DuPont, Syngenta and the other major genetic seed companies used similar forms of coercion, bribery and illegal tactics to spread their seeds from Poland to Indonesia and beyond. In Indonesia, Monsanto was forced to plead guilty to criminal charges of paying $50,000 in bribes to a senior Indonesian Government official to bypass controls on screening new genetically modified crops. Court records revealed that the bribe had been authorized in the US headquarters of Monsanto. Monsanto later was found guilty and forced to pay a fine.[30]

In Poland, Monsanto and the other major agribusiness corporations were illegally planting GMO seeds in a country with one of the richest soils in Europe. In Brazil, Monsanto was accused of illegally smuggling and planting large quantities of GMO soybean seed into the country. The Government finally lifted a ban on GMO plants in early 2005, stating it was futile to try to control the spread. The Gene Revolution was marching forward by all means possible.[31]

Killing Us Softly, Ever so Softly, Killing Us Softly With...

The clear strategy of Monsanto, Dow, DuPont and the Washington Government backing them was to introduce the GMO seeds in every corner of the globe, with priority on defenceless, highly indebted African and other developing countries, or countries like Poland and Ukraine where government controls were minimal and official corruption rampant.

Once planted, the seeds would spread rapidly across the land. At a later date, the GMO seed multinationals, using threats of WTO sanctions, would be in a position to dominate the seed supply of the major growing areas of the planet, to give or deny the means of life sustenance as they saw fit. In intelligence parlance, such a capacity was called the power of "strategic denial." A potential enemy or rival would be denied a strategic resource—energy or, in this case, food—or be threatened with denial, unless they agreed to certain policy demands by those controlling the resource.

A Very Special Kind of Corn

The question then became, how did this prospect map onto the long-term strategy of the Rockefeller Foundation, Ford Foundation and major figures in the US establishment for global population reduction? A possible answer was soon to be found.

In San Diego, a small, privately-owned biotech company, Epicyte, held a press conference in September 2001 to make an announcement about its work. Epicyte reported that they had successfully created the ultimate GMO crop—contraceptive corn. They had taken antibodies from women with a rare condition known as immune infertility, isolated the genes that regulated the manufacture of those infertility antibodies, and, using genetic engineering techniques, had inserted the genes into ordinary corn seeds used to produce corn plants.[32] "We have a hothouse filled with corn plants that make anti-sperm antibodies," boasted Epicyte President, Mitch Hein.[33]

At the time of this dramatic announcement, which went largely uncommented by the world's major media, Epicyte had concluded a strategic joint research and licensing agreement with Dow Chemical Company through Dow AgroSciences, one of the three agribusiness genetic seed giants in the US. The purpose of that joint venture, they announced at the time, was to combine Epicyte's technological breakthroughs with Dow AgroSciences' "strength in the genetic engineering of crops." Epicyte's product-candidate antibodies were being transformed in corn. Epicyte and the Dow organizations had agreed to a four year program to investigate factors affecting expression, stability and accumulation of antibodies in transgenic plants.[34] Epicyte had also signed a collaboration with Novartis Agriculture Discovery Institute (Syngenta) and with ReProtect LLC of Baltimore to develop other antibody-based microbicides for contraception.[35]

On October 6, 2002, CBS News reported that the United States Department of Agriculture, the same agency of the US Government that had been so vigorous in developing Terminator technology, had also financed 32 field trials around the country for growing drug and drug compounds in various crops. The US Government field trials included Epicyte's spermicidal corn technology. What was not revealed was that the USDA was also providing the field trial results to scientists at the US Department of Defense through one of their numerous biological research laboratories such as the Edgewood Chemical and Biological Center in Maryland.[36]

Previously, the production of antibodies for contraception purposes required costly facilities costing up to four hundred million dollars for ultra-sterile special fermentation conditions, using hamster ovarian bacteria as the antibody source. Epicyte claimed it needed only perhaps 100 acres of corn land to grow the special GMO spermicidal corn producing a vastly greater quantity of antibody for the spermicide at a cost of a mere few million dollars, a cost reduction of some 90%.[37]

At the time of their brief public announcement, which they presented as a contribution to the world "over-population" problem, Epicyte estimated the commercial availability of its spermicidal corn would come in 2006 or 2007. After the press release, the discussion of Epicyte's breakthrough in creating spermicidal corn which would kill human sperm vanished. The company itself was taken over in May 2004 by a private Pittsboro, North Carolina biotech company. Biolex thus acquired Epicyte Pharmaceutical.[38] Nothing more was heard in any media about the development of spermicidal corn and the theme vanished from view.

Rumors were that the research continued on a secret basis because of the politically explosive impact of a corn variety which, when consumed, would make human male sperm sterile. Mexican farmers were already in an uproar over the unauthorized spread of genetically engineered corn into the heart of the Mexican corn seed treasure in Oaxaca.[39]

It took little effort to imagine the impact were corn—which was the dietary staple of most Mexicans—to contain Epicyte's spermicidal antibodies. "Some spermicidal corn on the cob? Or perhaps a killer tortilla, mister?" Or what about that next bowl of corn flakes? The creator of the Kellogg's Corn Flakes company was also a founding patron of the American Eugenics Society almost a century before, along with John D. Rockefeller.

From Terminator Suicide Seeds to Spermicidal Corn

It was becoming clear why powerful elite circles of the United States, themselves enormously wealthy and largely untaxed thanks to Bush Administration tax cuts, backed the introduction of genetically modified seeds into the world food chain as a strategic priority. That elite included not only the Rockefeller and Ford foundations and most other foundations tied to the large private family fortunes of the wealthiest American families. It also included the US State Department, the National Security Council, the US Department of Agriculture, as well as the leading policy circles of the International Monetary Fund, the World Bank, along with agencies of the United Nations including WHO and FAO.

Tetanus, Rockefeller and the World Health Organization

The folks at the Rockefeller Foundation were deadly serious about wanting to solve the world hunger problem through the worldwide proliferation of GMO seeds and crops. Only their presumed method to do so aimed at a "supply side" solution rather than the "demand side." They were out to limit population by going after the human reproductive process itself.

For any skeptics who doubted their intent, they needed only to look at the foundation's work with the United Nations' World Health Organization in Mexico, Nicaragua, the Philippines and other poorer developing countries. The foundation had quietly funded a WHO program in "reproductive health," which had developed an innovative tetanus vaccine. It was no spur of the moment decision by the people at Rockefeller. Nor could they claim to be unaware of the true nature of their research funding. They had worked with WHO researchers since 1972 to develop a new double-whammy vaccine, at the same time during which the foundation had been funding research in other bio-technology areas including genetically engineered crops.[40]

In the early 1990's, according to a report from the Global Vaccine Institute, the WHO oversaw massive vaccination campaigns against tetanus in Nicaragua, Mexico and the Philippines. Comite Pro Vida de Mexico, a Roman Catholic lay organization, became suspicious of the motives behind the WHO program and decided to test numerous vials of the vaccine and found them to contain human Chorionic Gonadotrophin, or hCG. That was a curious component for a vaccine designed to protect people against lock-jaw arising from infection with rusty nail wounds or other contact with certain bacteria found in soil. The tetanus disease was indeed, also rather rare.[41]

It was also curious because hCG was a natural hormone needed to maintain a pregnancy. However, when combined with a tetanus toxoid carrier, it stimulated the formation of antibodies against hCG, rendering a woman incapable of maintaining a pregnancy, a form of concealed abortion. Similar reports of vaccines laced with hCG hormones came from the Philippines and Nicaragua.[42]

The Comite Pro Vida organization confirmed several other curious facts about the WHO vaccination program. The tetanus vaccine had been given only to women in the child-bearing ages between 15-45. It was not given to men or children.[43] Furthermore, it was usually given in a series of three vaccinations only months apart to insure that women had a high enough dosage of hCG, even though one tetanus injection held for at least ten years. The presence of hCG was a clear contamination of the vaccine. None of the women receiving the Tetanus hCG vaccine were told it contained an abortion agent. The WHO clearly intended it that way.

Pro Vida dug further and learned that the Rockefeller Foundation, working with John D. Rockefeller III's Population Council, the World Bank, the UN Development Program and the Ford Foundation, and others had been working with the WHO for 20 years to develop an anti-fertility vaccine using hCG with tetanus as well as with other vaccines.[44]

Among those "others" involved in funding the WHO research was a list which included the All India Institute of Medical Sciences, and a number of universities, including Uppsala in Sweden, Helsinki University, and Ohio State University. The list also included the US Government, through its National Institute of Child Health and Human Development, a part of National Institutes of Health (NIH). The latter US Government agency supplied the hCG hormone in some of the anti-fertility vaccine experiments.[45]

The respected British medical journal, *The Lancet,* in a June 11, 1988 article entitled "Clinical Trials of a WHO Birth Control Vaccine," confirmed the findings of the Comite Pro Vida de Mexico. Why a Tetanus Toxoid "Carrier"? Because the human body does not attack its own naturally occurring hormone hCG, the body has to be fooled into treating hCG as an invading enemy in order to develop a successful anti-fertility vaccine utilizing hCG anti-bodies, according G.P. Talwar, one of the scientists involved.[46]

By mid-1993, the WHO had spent a total of $365 million of its scarce research funds on what it euphemistically dubbed "repro-ductive health," including research on implanting hCG into tetanus vaccines. WHO officials declined to explain why women they had vaccinated had developed anti-hCG antibodies.[47]

They dismissed the findings of Pro Vida by claiming the charges were coming from "Right to Life and Catholic sources;" as if that should indicate some fatal bias. If you can't deny the message, at least try to discredit the messenger.

When four additional vials of the tetanus vaccine used on women in the Philippines were sent to St. Luke's Lutheran Medical Center in Manila, where all four tested positive for hCG, the officials at WHO shifted. The WHO now claimed the hCG had come from the manufacturing process.

The vaccine had been produced by Connaught Laboratories Ltd of Canada, Intervex and CSL Laboratories of Australia. Connaught, one of the world's largest producers of vaccines, was part of the French pharmaceutical Rhône Poulenc group. Among other research projects, Connaught was engaged in producing a genetically engineered version of the Human Immunodeficiency Virus (HIV).

Population reduction and genetically engineered crops were clearly part of the same broad strategy: the drastic reduction of the world's population. It was in fact a sophisticated form of what the Pentagon termed biological warfare, promulgated under the name of "solving the world hunger problem."

The Hidden GMO Agenda Emerges
The US and UK governments' relentless backing for the global spread of genetically modified seeds was in fact the implementation of a decades long policy of the Rockefeller Foundation since the 1930's, when it funded Nazi eugenics research—i.e. mass-scale pop-ulation reduction, and control of darker-skinned races by an Anglo-

Saxon white elite. As some of these circles saw it, war as a means of population reduction was costly and not that efficient.

In 1925, Britain's Winston Churchill, a robust racist, commented favorably on the potential for biological warfare, writing about the desirability of the government being able to produce "pestilences methodically prepared and deliberately launched upon man and beast.... Blight to destroy crops. Anthrax to slay horses and cattle...." And that was in 1925.[48]

Reflecting the discussion in US senior military circles, Lt. Col. Robert P. Kadlec, USAF, of the College of Aerospace Doctrine, Research and Education, discussed in a book written in the 1990's, *Battlefield of the Future*, the biowarfare potential of genetically engineered crops. He referred to GMO-based biological weapons as "cost-effective" weapons of mass destruction. He wrote that, "Compared with other mass destruction weapons, biological weapons are cheap. A recent Office of Technology Assessment report paces the cost of a BW (biological weapon) arsenal as low as $10 million ... in stark contrast to a low-estimate of $200 million for developing a single nuclear weapon."[49]

Kadlec then went on to state that, "Using biological weapons under the cover of an endemic or natural disease occurrence provides an attacker the potential for plausible denial. In this context, biological weapons offer greater possibilities for use than do nuclear weapons."[50]

The biological weapons and genetic engineering research project, Sunshine Project, reported that "researchers in the USA, UK, Russia and Germany have genetically engineered biological weapons agents, building new deadly strains.... Genetic engineering can be used to broaden the classical bio-weapons arsenal ... bacteria can not only be made resistant to antibiotics or vaccines, they can also be made more toxic, harder to detect...."[51]

Back in the 1980's around the time the Rockefeller Foundation launched its major genetic engineering rice project, the start of the Gene Revolution, the US Pentagon quietly initiated military applications of biotechnology. Citing the Russian threat, US military researchers, in highly classified research, began using the new genetic engineering techniques. Among the projects researched was a genetically modified refined opium-like substance, whose minute presence induced sleep, anxiety, submissiveness or temporary blindness.

Significantly, in the context of Terminator, GMO spermicides and other developments of the Gene Revolution, the Bush Administration rejected a ban on further bio-weapons development, and at the same time refused to accept the Kyoto Protocol on global warming and CO_2 emissions.[52] The bio-weapons protocol was a major issue among the list of things the new administration in Washington unilaterally rejected. The media dutifully turned its focus, however, to the Bush rejection of Kyoto, largely ignoring the Administration's significant refusal to cooperate on banning biological and toxic weapons.

In one of his first acts after taking office in January 2001, Bush announced that he refused to support a legally binding Biological and Toxic Weapons Protocol (BTWC), leading to collapse of those international talks. Little reason was given. A 2004 study by the British Medical Association concluded that the world was perhaps only a few years away from "terrifying biological weapons capable of killing only people of specific ethnic groups," citing advances in "genetic weapons technology."[53]

"We're tempted to say that nobody in their right mind would ever use these things," remarked Stanford University biophysicist, Professor Steven Block, a man with years of personal experience with classified Pentagon and Government biological research. "But," Block added, "not everybody is in their right mind...."[54]

Notes

1. Melvin John Oliver et al., United States Patent, *Control of Plant Gene Expression*, Patent no. 5,723,765, 3 March 1998, detailed description of the invention in http://patft1.uspto.gov.

2. Geri Guidetti, "Seed Terminator and Mega-Merger Threaten Food and Freedom, Food Supply Update: June 5, 1998", *The Ark Institute*, http://www.arkinstitute.com/98/up0606.htm, 5 June 1998.

3. Melvin John Oliver, *op. cit.*

4. Hugh Warwick, *Syngenta: Switching off Farmers' Rights?*, Genetics Forum, Bern, October 2000, http://www.mindfully.org/GE/Syngenta-Switching-Off-Rights.htm#exec.

5. *Ibid.*

6. Cooperative Research, *March 3, 1998: Patent on Terminator Seeds Granted*, http://www.cooperativeresearch.org/entity.jsp?entity=us_department_of_agriculture.

7. Zac Hanley and Kieran Elborough, "Re-emerging Biotechnologies: Rehabilitating the Terminator", *ISB News Report*, June 2002, p. 1.

8. United States Department of Agriculture Agricultural Research Service, *op. cit.*, 29 March 2004.

9. Martha L. Crouch, *How the Terminator Terminates*, lists.ibiblio.org/pipermail/permaculture/1999-January/005941.html, Indiana University, Bloomington, Indiana, 1998.

10. US Patent and Technology Office, *USPTO Patent Database*, 3 March 1998.

11. USA Today, "The Seeds of Warning for Biotech Companies", *USA Today*, 10D, 29 June 1999.

12. Robert B. Shapiro, *Open Letter From Monsanto CEO Robert B. Shapiro To Rockefeller Foundation President Gordon Conway and Others*, http://www.monsanto.co.uk/news/ukshowlib.phtml?uid=9949, 4 October 1999.

13. *Ibid.*

14. *Ibid.*

15. Willard Phelps, USDA spokesman, *Interview with RAFI* (now ETC), 10 March 1998, cited in http://www.cropchoice.com/leadstry7f4c.html?recid=694.

16. Melvin J. Oliver, USDA molecular biologist and primary inventor of the technology, quoted in *RAFI Communique*, March 1998.

17. GRAIN, "Confronting Contamination: 5 Reasons to Reject Co-existence, Contamination in Argentina and Brazil pays off for Monsanto", *Seedling*, http://www.grain.org/seedling, April 2004.

18. RAFI, 25 February 2000, quoted in Jaan Suurküla, *Problems with Genetically Engineered Food Archive: RAFI Says Terminator Seeds on Fast Track*, http://www.psrast.org/probobstarch.htm.

19. United States Department of Agriculture Agricultural Research Service, *op. cit.*, 29 March 2004.

20. *Ibid.*

21. Ashok B. Sharma, "US Aid Agencies Instructed to Report Anti-GM Nations to USAID", *The Financial Express* (India), http://www.mindfully.org/GE/2003/USAID-Report-AntiGM14jan03.htm, 14 January 2003.

22. European Commission, *WTO and Agriculture: European Commission Proposes More Market Opening, Less Trade Distorting Support and a Radically Better Deal for Developing Countries*, Press release, 16 December 2002.

23. Ashok B. Sharma, *op. cit.*

24. John Vidal, "US Dumping Unsold GMO Food on Africa", *The Guardian*, 7 October 2002.

25. *Ibid.*

26. *Ibid.*

27. Charles Benbrook, quoted in "Southern Africa's Food Aid Crisis Shamelessly Engineered to Score 'Political Points' Says Leading US Agronomist", *Norfolk Genetic Information Network*, ngin.tripod.com/270902a.htm, 27 September 2002.

28. Mark Townsend, "Blair Urges Crackdown on Third World Profiteering", *The Observer*, 1 September 2002.

29. BBC News, *Bush: Africa Hostage to GM Fears*, 22 May 2003, http://news.bbc.co.uk/2/hi/americas/3050855.stm.

30. Jonathan Birchall, "Indonesia: Monsanto Agrees to US $1.5 Million Over Crop Bribe", *Financial Times* (London), 7 January 2005.

31. Andrew Hay, "Environmentalists Fear Brazil's Lifting of GMO Ban," *Reuters*, 7 March 2005.

32. Robin McKie, "GMO Corn Set to Stop Man Spreading His Seed", *The Observer*, 9 September 2001. McKie writes, "The pregnancy prevention plants are the handiwork of the San Diego biotechnology company Epicyte, where researchers have discovered a rare class of human antibodies that attack sperm. By isolating the genes that regulate the manufacture of these antibodies, and by putting them in corn plants, the company has created tiny horticultural factories that make con-

traceptives... Contraceptive corn is based on research on the rare condition, immune infertility, in which a woman makes antibodies that attack sperm...Essentially, the antibodies are attracted to surface receptors on the sperm," said Hein. "They latch on and make each sperm so heavy it cannot move forward. It just shakes about as if it was doing the lambada."

33. *Ibid.*

34. PRNewswire, *Dow, Epicyte Enter Research, Licensing Agreement*, 5 September 2000.

35. "Epicyte: Company of the Month", *The San Diego Biotech Journal*, June 2001, http://www.biotechjournal.com/Journal/Jun2001/juneartA2001.pdf.

36. Wyatt Andrews, "In Coming Harvests, Farm-aceutical Corn", *CBS News*, http://www.muhammadfarms.com/News-Oct6-12-2002.htm, 8 October 2002.

37. The San Diego Biotech Journal, *op. cit.* See also Business Wire, *Epicyte Receives SBIR Grant to Fund HPV Antibody Development; Marks Fifth Grant for Epicyte to Develop Sexual Health Products*, 5 June 2001.

38. *Biolex Acquires San Diego Based Epicyte Pharmaceutical*, Company Press Release, 6 May 2004. See http://www.biolex.com and http://www.epicyte.com

39. S'ra DeSantis, "Mexico: Genetically Modified Organisms Threaten Indigenous Corn", *Z Magazine*, July-August 2002.

40. "Clinical Trials of a WHO Birth Control Vaccine", *The Lancet*, 11 June 1988.

41. James A. Miller, "Are New Vaccines Laced With Birth-Control Drugs?", *HLI Reports*, Human Life International, Gaithersburg, Maryland; June-July 1995.

42. *Ibid.*

43. *Ibid.*

44. *Ibid.* The author cites details of official WHO articles on birth control vaccines including, "Vaccines for Fertility Regulation," Chapter 11, pp. 177-198, *Research in Human Reproduction, Biennial Report*, 1986-1987, WHO Special Programme of Research, Development and Research Training in Human Reproduction, WHO, Geneva, 1988.

45. James A. Miller, *op. cit.*

46. W.R. Jones, et al., "Phase 1 Clinical Trials of a World Health Organisation Birth Control Vaccine", *The Lancet*, 11 June 1988, pp. 1295-1298. The authors write, "A birth control vaccine incorporating a synthetic peptide antigen representing the aminoacid sequence...of human chorionic gonadotropin (hCG-beta) was submitted to a phase 1 clinical trial. Thirty surgically sterilised female volunteers, divided into five equal groups for different vaccine doses, received two intramuscular injections six weeks apart. Over a six-month follow-up...potentially

contraceptive levels of antibodies to hCG developed in all subjects. In the highest vaccine dose group, the results gave promise of a contraceptive effect of six months' duration." Also, G.P. Talwar, et al., "Prospects of an Anti-hCG Vaccine Inducing Antibodies of High Affinity…", *Reproductive Technology*, 1989, Elsevier Science Publishers, 1990, Amsterdam, New York, p. 231.

47. James.A. Miller, *op. cit.* Also World Health Organization, "Challenges in Reproductive Health Research", *Biennial Report, 1992-1993*, Geneva, 1994, p. 186.

48. Winston Churchill, quoted in Robert Harris and Jeremy Paxman, *A Higher Form of Killing*, Noonday Press, New York, 1982. See also George Rosie, "Churchill's Anthrax Bombs: UK Planned to Wipe out Germany with Anthrax", *Sunday Herald*, London, 14 October 2001.

49. Robert P.Kadlec, *Biological Weapons for Waging Economic Warfare and Twenty-First Century Germ Warfare*, http://www.airpower.maxwell.af.mil/airchronicles/battle/chp10.html, and …chp9.html.

50. *Ibid.*

51. Sunshine: Biological Weapons and Genetical Engineering, *Genetic Engineering is Regularly Used to Produce Lethal Bacteria*, http://www.sunshine-project.org/bwintro/gebw.html.

52. *Ibid.*

53. Helen Nugent, "Gene Wars Only a Few Years Away, Say Doctors", *London Times*, 26 October 2004.

54. Steven Block, quoted in Mark Shwartz, "Biological Warfare Emerges as 21st-Century Threat", *Stanford Report*, news-service.stanford.edu/news/2001/january17/bioterror-117.html, 11 January 2001.

CHAPTER 13
Avian Flu Panic and GMO Chickens

The President Helps out a Friend

On November 1, 2005, President George W. Bush went to the National Institutes of Health in Bethesda, Maryland to hold an unusually high profile press conference to announce a 381-page plan officially called the Pandemic Influenza Strategic Plan. It was in many respects as unusual and significant as the President's May 2003 press conference where he declared his intent to file WTO action to break the European Union moratorium on GMO.

The NIH press conference was no ordinary Bush photo opportunity. This one was meant to be a big event. The President was surrounded by almost half his cabinet, including Secretary of State Condoleezza Rice, joined by the Secretaries of Homeland Security, Agriculture, Health & Human Services, Transportation and, interestingly enough, Veteran Affairs. And just to underscore that this was a big deal, the White House invited the Director-General of the World Health Organization, who flew in from Geneva, Switzerland for the occasion.

The President began his remarks, "at this moment, there is no pandemic influenza in the United States or the world. But if history is our guide, there is reason to be concerned. In the last century, our country and the world have been hit by three influenza pandemics—and viruses from birds contributed to all of them ...."

Bush spoke about an imminent danger to the American people: "Scientists and doctors cannot tell us where or when the next pandemic will strike, or how severe it will be, but most agree: at some point, we are likely to face another pandemic. And the scientific community is increasingly concerned by a new influenza virus known as H5N1—or avian flu ...."

The President went on to warn:

> At this point, we do not have evidence that a pandemic is imminent. Most of the people in Southeast Asia who got sick were handling infected birds. And while the avian flu virus has spread from Asia to Europe, there are no reports of infected birds, animals, or people in the United States. Even if the virus does eventually appear on our shores in birds, that does not mean people in our country will be infected. Avian flu is still primarily an animal disease. And as of now, unless people come into direct, sustained contact with infected birds, it is unlikely they will come down with avian flu.[1]

Bush then called on Congress to immediately pass a new bill with $7.1 billion in emergency funding to prepare for that possible danger. The speech was an exercise in the Administration's "pre-emptive war," this time against avian flu. As with the other pre-emptive wars, it followed a multiple agenda.

Prominent among the President's list of emergency measures was a call for Congress to appropriate another $1 billion explicitly for a drug developed in California called Tamiflu. The drug was being heavily promoted by Washington and the WHO as the only available medicine to reduce symptoms of general or seasonal influenza, which also "possibly" might reduce symptoms of avian flu. The large Swiss pharmaceutical firm, Roche, held the sole license to manufacture Tamiflu. With growing scare stories in US and international media warning of the deadly new H5N1 strain of Avian Flu virus and the "high" risk of human-to-human contamination, order books at Roche were backed up for months.

What President Bush neglected to say was that Tamiflu had been developed and patented by a California biotech firm, Gilead Science Inc., a listed US stock company which preferred to maintain a low profile in the context of growing interest in Tamiflu. That might have been because in 1997, before he became US Secretary of Defense in the Bush Administration, the President's close friend, Donald H. Rumsfeld, had been Chairman of the Board of Gilead Science Inc. He had remained there until early 2001 when he became Defense Secretary. Rumsfeld had been on the Gilead board since 1988 according to a January 3, 1997 company press release.[2]

In November 2004, while Rumsfeld was Defense Secretary, his Assistant Secretary for Health Affairs issued a directive regarding Avian Flu. The document stated that, "… oseltamivir (Tamiflu) will be used to prevent and treat illness. There is evidence that H5N1 is sensitive to oseltamivir. However, its supply is extremely limited worldwide, and its use will be prioritized."[3] That 2004 Pentagon directive made a significant contribution to panic buying of Tamiflu by governments around the world.

Unconfirmed reports were that while Rumsfeld was Secretary of Defense, he also purchased additional stock in his former company, Gilead Science, worth $18 million, making him one of the largest— if not the largest—Gilead stock owners. He stood to make a fortune on royalties and on the rising stock price for Gilead, as a panicked world population scrambled to buy a drug whose capacity to cure the alleged avian flu was still uncertain.[4]

This phenomenon suggested a parallel with the corruption of Halliburton Corporation, whose former CEO was Vice President Dick Cheney. Cheney's Halliburton had gotten billions of dollars worth of US construction contracts in Iraq and elsewhere.[5]

Was the avian flu scare another Pentagon hoax, whose ultimate aim was unknown?

Kissinger and Biological Warfare

Back in the mid-1970's, acting as National Security Advisor (NSA) under Richard Nixon, Nelson Rockefeller's protégé Henry Kissinger oversaw foreign policy, including his NSSM 200 project, the top secret Third World population reduction strategy for the US, Britain, Germany, and other NATO allies. According to the US Congressional Record of 1975, Kissinger selected to have the Central Intelligence Agency (CIA) develop biological weapons.[6] Among these new man-made biological weapons were germs far deadlier than the avian flu.[7]

By 1968, when Kissinger requested and received updated intelligence on useful "synthetic biological agents" for germ warfare and population control, mutant recombinant flu viruses had just been engineered by US Government Special Virus Cancer Program researchers. During this program, influenza and para-influenza viruses were recombined with quick-acting leukemia viruses to deliver weapons that potentially spread cancer, like the flu, by sneezing. These researchers also amassed avian cancer (sarcoma) viruses and inoculated them into humans and monkeys to determine their carcinogenicity, according to AIDS researcher, Dr. Leonard Horowitz.[8]

In related efforts, US government researchers used radiation to enhance the cancer-causing potential of the avian virus. Those incredible scientific realities were officially censored. The sudden emergence of a global scare over a supposedly deadly strain of Avian Flu virus in 2003 had to be treated with more than a little suspicion.

Agribusiness Gains in Avian Flu Scare

Not only was Defense Secretary Rumsfeld a direct benefactor of US, UK and other governments' stockpiling of his Tamiflu, the avian flu scare was also being used to advance the global domination of agribusiness and poultry factory farms along the model of the Arkansas-based Tyson Foods Inc.

Curiously enough indeed, the huge, unsanitary and overcrowded factory chicken farms of the global agribusiness giants were not those being scrutinized as a possible incubator or source of H5N1 or other diseases. Instead, the small family-run chicken farmers especially in Asia, with at most perhaps 10 to 20 chickens, were those who stood to lose in the Bird Flu hysteria.

The major US chicken factories such as Tyson Foods, Perdue Farms and ConAgra Poultry made a propaganda campaign, falsely claiming that, unlike those of free-running Asian chicken farms, their chickens were "safer" because they were raised in closed facilities.

As an integral part of the Harvard-run agribusiness vertical integration project of Professors John Davis and Ray Goldberg, the US poultry industry became one of the first targets for industrialization or "factory farming."[9]

The industrialization of chicken-raising and slaughtering in the USA had progressed to the point that by 2003 when the first cases of H5N1 avian flu virus were reported from Asia, five giant multinational agribusiness companies dominated the production and processing of chicken meat in the United States. Indeed, according to trade source, *WATT Poultry USA*, as of 2003 the five companies held overwhelming domination of the US poultry production, all of them vertically integrated.[10]

The five companies were Tyson Foods, the largest in the world; Gold Kist Inc., Pilgrim's Pride, ConAgra Poultry, and Perdue Farms. In January 2007, Pilgrim's Pride bought Gold Kist, creating the largest chicken agribusiness giant. Together, the five accounted for over 370 million pounds, per week, of ready-to-cook chicken, corresponding to some 56% of all ready-to-eat poultry produced in the USA. The US chicken factory farms produced almost 9 billion "broiler" or meat chickens in 2005, or 48 billion pounds of chicken meat. The State of Arkansas, home of Tyson Foods, produced 6,314,000,000 pounds of that chicken meat.[11]

They produced chicken meat in atrocious health and safety conditions. In January 2005, a US Government Accountability Office (GAO) report to the US Senate, "Safety in the Meat and Poultry Industry," concluded that US meat and poultry processing plants had "one of the highest rates of injury and illness of any industry." They cited exposure to "dangerous chemicals, blood, fecal matter, exacerbated by poor ventilation and often extreme temperatures." Workers typically faced hazardous conditions, loud noise, must work in narrow confines with sharp tools and dangerous machinery.[12]

Another report from VivaUSA, a non-profit organization investigating conditions in US factory farms, noted that, "thanks to genetic selection, feed, and being prevented from moving or getting any exercise on factory farms, chickens now grow to be much larger and to grow more quickly than ever before." They cite a USDA study which noted that "in the 1940's broilers required 12 weeks to reach market weight (4.4 pounds), whereas, due to the unnatural elements of industrialized production methods, now they reach that weight and are killed at just six weeks of age." [13]

The use of the growth boosters created major health problems in the huge factory farm concentrations. Because of hormone and vaccine injections used to speed growth, muscle growth outstripped bone development and the chickens typically had leg and skeletal disorders that affected their ability to walk. Unable to walk, they had to sit in poor-quality litter, creating breast blisters or hock burns. Chicken organs were unable to keep up with their hyper growth rates, causing hearts or lungs to fail or malfunction, creating excess fluids in their bodies, or death. [14]

Under special exemptions in US law, chickens were excluded from the protections of the federal Animal Welfare Act. The federal government set no rules or standards for how chickens should be housed, fed, or treated on farms. According to a growing number of animal health experts, factory farming, rather than the small free roaming chicken operations of Asia, was the real source of horrendous new diseases and viruses such as H5N1.

A World GMO Chicken?

Alone, Tyson Foods processed 155 million pounds of chicken a week, almost three times the production of its nearest rival. Tyson made over $26 billion a year in revenue in 2006. During the peak of the bird flu scare, the Quarter ending September 30, 2005, Tyson Foods" earnings rose by 49%. Its profit margin in chickens grew by 40%. [15] Tyson Foods and the small international cartel of poultry agribusiness firms stood to gain from the avian flu scare.

The giant American chicken processors were out to globalize world chicken production by the turn of the Millennium. The avian flu seemed a gift from Heaven, or Hell, sent precisely for that task. One clear target for those companies was the huge Asian poultry market. Were Asian governments forced via WHO and international pressure to force farmers to cage chickens, small farmers would be bankrupted and large agribusiness firms like Tyson Foods or the Thailand-based CP Group would thrive.

In a detailed report issued in February 2006, GRAIN, an organization dealing with GMO issues, revealed that the Thai-based CP Group and other chicken factory farms "were present nearly everywhere bird flu has broken out."[16] The outbreaks which had been traced as far away as Turkish Anatolia, Bulgaria and Croatia by early 2006 all followed the transportation routes by air or rail of processed poultry from CP Group operations in China, Thailand, Cambodia or elsewhere in Asia where mass crowding and unsanitary closed conditions provided ideal breeding conditions for the outbreak of disease.

The GRAIN report noted:

> The transformation of poultry production in Asia in recent decades is staggering. In the Southeast Asian countries where most of the bird flu outbreaks are concentrated—Thailand, Indonesia, and Viet Nam—production jumped eightfold in just 30 years, from around 300,000 metric tonnes (mt) of chicken meat in 1971 to 2,440,000 mt in 2001. China's production of chicken tripled during the 1990s to over 9 million metric tons per year.[17]

Practically all of this new poultry production has happened on factory farms concentrated outside of major cities and integrated into transnational production systems. This is the ideal breeding ground for highly-pathogenic bird flu—like the H5N1 strain threatening to explode into a human flu pandemic.

A report by a Canadian organization, *Beyond Factory Farming*, described the transmission likely pathways from the giant industrialized chicken centers:

> In Thailand, China and Vietnam there is a highly developed industrial poultry industry which has expanded dramatically in the past decade. The large poultry companies raise millions of birds, hatch chicks to supply other intensive poultry operations, export live birds and eggs to countries such as Nigeria (where the first Highly Pathogenic Avian Influenza outbreak in Africa was recently reported)

and produce and export feed which often includes "litter" (i.e., manure) in the ingredients.

[...]

Manure that may contain live virus is spread on surrounding farmland, or exported as fertilizer, and through run-off may end up in surface waters where wild birds feed and rest. Chicken manure is even found in fish farm feed formulations where it is introduced directly into the aquatic environment. Wild birds and poultry that have fallen victim to HPAI in Asia, Turkey and Nigeria appear to have been directly exposed to HPAI virus originating in the factory farm system. In Asia, a flock of wild ducks died from HPAI—after having come into contact with the disease at a remote lake where a fish farm used feed pellets made from poultry litter from a factory farm. In Turkey a massive cull of backyard flocks—and the deaths of three children—took place after a nearby factory farm sold sick and dying birds to local peasants at cut rate prices. Nigeria has a large and poorly regulated factory poultry production sector which is supplied with chicks from factory farms in China.[18]

As experts on migratory bird flight pointed out, birds migrate in late fall from the Northern Hemisphere to southern, sunnier climates for winter. Bird flu outbreaks followed an East-West route, not North-South. Officials at WHO and the US Government's Centers for Disease Control conveniently omitted that salient fact as they spread fear of free-flying birds.[19]

CP Group of Thailand, Asia's largest poultry factory farming agribusiness group, was no mom-and-pop operation. By 2005, it had operations in more than 20 countries, including China, where, under the name Chia Tai Group, it employed 80,000 people.[20]

Group patriarch, Dhanin Chearavanont, a billionaire with a penchant for cock fighting and yachts, was hardly a struggling third world businessman. He started in 1964 when he learned the concept of vertical integration from Arbor Acres Farm of Connecticut in the United States, at the time the world's largest chicken factory, financed by Nelson Rockefeller. Chearavanont was business partner with among others, Neil Bush, brother of George W. Bush, and his own executive Vice President Sarasin Viraphol, former Thai Deputy

Secretary of Foreign Affairs, was chosen to sit on David Rockefeller's elite Trilateral Commission.[21]

By early 2006, it seemed clear that the five or six giant poultry agribusiness multinationals, five US-based and one Thai-based and White House-connected, were moving to industrialize the majority of world chicken production, the main meat protein source for much of the planet, especially in Asia.

One little-noted research project in England gave a clue as to what the subsequent phase of the globalization of chicken production would be. Once it would be produced in massive factory farm installations worldwide, the world chicken population would be an easy target for the creation of the first GMO animal population.[22]

Amid reports of spreading bird flu from Asia across to Europe, the London *Times* noted in its October 29, 2005 edition, that a very active research project at Scotland's Roslin Institute, operating in collaboration with Laurence Tiley, Professor of Virology at Cambridge University, was on the brink of genetically engineering chickens to produce birds resistant to the lethal strains of the H5N1 virus. The new "transgenic chickens" would have small pieces of genetic material inserted into chicken eggs to allegedly make the chickens H5N1 resistant.[23]

Roslin Institute had earlier contracted with a Florida biotech company, Viragen, for the rights to commercialize Avian Transgenic Technology, a method in which flocks of specially produced transgenic chickens would lay virtually unlimited numbers of eggs expressing high volumes of the target drug in the egg whites.[24] Roslin had first captured world headlines with their creation of "Dolly the Sheep."

Tiley was buoyant about the prospects for transforming the world chicken population into GMO birds. He told the *Times* that "once we have regulatory approval, we believe it will only take between four and five years to breed enough chickens to replace the entire world (chicken) population."

Within the space of little more than two decades, GMO science had enabled a small handful of private global agribusiness companies—three of them American-based—to secure a major foothold and patent rights to world production of such essential feed grains as rice, corn, soybeans and soon wheat. By 2006, riding the fear of an Avian Flu human pandemic, the GMO or Gene Revolution players were clearly aiming to conquer the world's most important source of meat protein, poultry.

Soon the next piece in the global control over man's food chain was executed. It played out on a quiet August day in Scott, Mississippi. The implications were staggering. Terminator was about to come into the control of the world's largest GMO agribusiness seed giant.

Notes

1. George W. Bush, *President Outlines Pandemic Influenza Preparations and Response,* Washington D.C., NIH, http://www.whitehouse.gov/news/releases/2005/11/20051101-1.html, 1 November 2005.

2. Gilead Sciences, *Donald H. Rumsfeld Named Chairman of Gilead Sciences,* Press Release, Foster City, CA., 3 January 1997, http://www.gilead.com/wt/sec/pr_933190157/.

3. William Winkenwerder Jr., *Department of Defense Guidance for Preparation and Response to an Influenza Pandemic caused by the Bird Flu (Avian Influenza),* US Department of Defense, http://www.geis.fhp.osd.mil/GEIS/SurveillanceActivities/Influenza/DoD_Flu_Plan_040921.pdf., 21 September 2004.

4. F. William Engdahl, "Is Avian Flu another Pentagon Hoax?", *Global Research,* http://www.globalresearch.ca/index.php?context=viewArticle&code=%20EN20051030&articleId=1169, 30 October 2005.

5. Rep. Henry A Waxman, *Halliburton's Iraq Contracts Now Worth over $10 Billion,* Committee on Government Reform, US House of Representatives, Washington, D.C., Fact Sheet, 9 December 2004, http://www.truthout.org/mm_01/5.120904A-1.pdf.

6. Leonard G. Horowitz, "Emerging Viruses: AIDS & Ebola, Nature, Accident or Intentional?", *Sandpoint,* Tetrahedron Publishing Group, Idaho, 2001, pp. 275-288.

7. *Ibid.,* p. 411.

8. *Ibid.,* pp. 410-411.

9. Ira Wolfert, "Chickens: Cheaper by the Mission", *The Reader's Digest,* February 1968.

10. WATT Poultry USA, *WATT Poultry USA's Rankings,* October 2006.

11. Viva! USA, *Chicken/Broiler Industry Media Briefing,* http://www.vivausa.org/campaigns/chickens/media.html, 2005.

12. United States Government Accountability Office, *Safety in the Meat and Poultry Industry, While Improving, Could Be Further Strengthened,* Washington, D.C., January 2005, GAO-05-96.

13. Viva! USA, *op. cit.*

14. USDA, *Animal Welfare Issues Compendium. A Collection of 14 Discussion Papers,* September 1997. http://warp.nal.usda.gov/awic/pubs/97issues.htm. Accessed on 30 September 2005.

15. Tyson Foods, Inc., *Annual Report, 2006,* http://www.tyson.com/Corporate/.

16. GRAIN, *Fowl Play: The Poultry Industry's Central Role in the Bird Flu Crisis*, http://www.grain.org/go/birdflu, February 2006.

17. *Ibid.*

18. Beyond Factory Farming Coalition, *Fact Sheet: Avian Flu*, http://www.beyondfactoryfarming.org/documents/Avian_Flu_Fact_Sheet.pdf. Cited in GRAIN, *op. cit.* See also, World Health Organization, *Bird Droppings Prime Origin of Bird Flu*, 17 January 2004, Geneva.

19. Walter Sontag, "Der Fluch der Vögel", *Wiener Zeitung*, 5 November 2005.

20. Details of CP Group can be found on the company's website, http://www.cpgroup.cn, and *Time Asia* magazine, "The Families that Own Asia", http://www.time.com/time/asia/covers/501040223/chearavanont.html.

21. Trilateral Commission, *The 2005 Trilateral Commission Membership List*, New York, May 2005.

22. Roslin Institute, *Research Reviews, Practical Environmental Enrichment to Improve Poultry Welfare*, pp. 55-60, http://www.roslin.ac.uk/research/hostResponse.php.

23. Mark Henderson, "Scientists Aim to Beat Flu with Genetically Modified Chickens", *The Times*, 29 October 2005.

24. Virugen, http://www.viragen.com/aviantransgenicbio.htm.

CHAPTER 14
Genetic Armageddon:
Terminator and Patents on Pigs

Monsanto Finally Takes Delta & Pine Land

On a Summer day in August 2006, as much of the world was lost in vacation distractions, a corporate acquisition took place which was to set the stage for the final phase of the Rockefeller Foundation's decades-long dream of controlling the human species.

On August 15, 2006, Monsanto Corporation, the Goliath of GMO agribusiness, announced that it had made a new bid to take ownership of Delta & Pine Land of Scott, Mississippi. The disclosed purchase price was $1.5 billion in cash.[1] Unlike when it had tried the same ploy in 1999 and was forced to back down by a storm of public protest, this time the takeover went almost unnoticed. The timing of the second takeover bid by Monsanto coincided with statements by Delta & Pine Land as to when they would be ready to commercialize Terminator.

The NGOs which had drawn attention to the Terminator issue in 1999 were hardly to be heard beyond a brief perfunctory press release or two. The Major US and international media ran the story under headlines similar to that in the New York Times: "Monsanto Buys Delta and Pine Land, Top Supplier of Cotton Seeds in US."[2]

Only far down in the last sentence of the article did the Times even note that Delta & Pine Land held "a controversial genetic engineering technology that makes sterile seeds."

The once-vocal public voice of the Rockefeller Foundation was this time silent. In 1999, Foundation President, Gordon Conway, a passionate advocate of what he even dubbed the Gene Revolution, made a concerted intervention. He personally argued with the board of Monsanto that Delta & Pine Land's Terminator patents, in the hands of a giant GMO company like Monsanto, risked a public revolution against spread of GMO.[3]

This time around, the influential Rockefeller Foundation did not even bother to issue a press release opposing the planned second try to capture Terminator rights by Monsanto. Foundation Press Spokesman Peter Costiglio, in reply to a public question tersely replied: "We don't have a statement to share with you …. The Rockefeller Foundation still opposes the use of Terminator technology in developing (*sic*) countries."[4] They declined to oppose Terminator universally, despite the fact that farmer-saved seeds are a major factor throughout the industrialized world as well.

The general yawn of reaction to the second Terminator takeover bid by Monsanto tended to confirm the fears of skeptics who warned in 1999 that Monsanto's Terminator dreams had anything but "terminated." They were only dormant until public opposition had weakened.

Wall Street stock traders greeted the takeover with jubilation and the price of stock of D&PL went ballistic from $27 a share in early August to over $40, a jump of more than 50% in days.

Monsanto crop-biotech competitors DuPont and Swiss-based Syngenta, both in a bitter battle to gain market share from Monsanto, lobbied for Justice Department involvement to block the D&PL takeover by rival Monsanto. DuPont said in a statement, "we have serious concerns about the impact that it would have on farmers, the agriculture industry and ultimately consumers." Their "concern" appeared to be more directed at the staggering implications of Monsanto now controlling world rights for Terminator, a process aided and abetted by the US Government, through the cooperation of the US Department of Agriculture in Delta & Pine Land's Terminator research.[5]

EU Patent Office Approves Terminator

In the intervening seven years since the first attempt by Monsanto to acquire Delta & Pine Land and its global Terminator patent rights, D&PL had not been idle. It had aggressively and success-fully extended its patent rights on GURTs. In October 2005 Delta & Pine Land together with the US Department of Agriculture won a major new patent on its Terminator technology from the European Union's European Patent Office, Patent no. EP775212B. The patent would cover all 25 nations in the European Union from Germany to Poland and Italy to France, some of the world's most abundant food-producing regions.

Several days later D&PL and the US Government also secured patent protection for its Terminator technology in Canada under CA 2196410. The advance of Terminator technology to global com-mercialization had hardly ceased despite the de facto worldwide UN ban imposed years before.[6]

The advent of GMO patented seeds on a commercial scale in the early 1990's had allowed companies like Monsanto, DuPont and Dow AgroSciences to go from supplying agriculture chemical herbicides like Roundup, to patenting genetically altered seeds for basic farm crops like corn, rice, soybeans or wheat. For almost a quarter century, since 1983, the US Government had quietly been working to perfect a genetically engineered technique whereby farmers would be forced to turn to their seed supplier each har-vest to get new seeds.

At the Fourth Meeting of the Working Group of the international Convention on Biological Diversity of the United Nations Environ-ment Program in Granada in January 2006, a group of indigenous farmers from Peru filed a submission on their concerns over possible introduction of Terminator seed technology:

> As traditional indigenous farmers we are united to defend our liveli-hoods which are dependant on seeds obtained from the harvest as

a principal source of seed to be used in subsequent agricultural cycles. This tradition of seed conservation underpins Andean and Amazonian biodiversity and livelihood strategies, the traditional knowledge and innovation systems customarily administered by indigenous women who have made such biodiversity and livelihood strategies possible and indigenous cultural and spiritual values that honor fertility and continuity of life.

Their petition to ban Terminator internationally argued several points cogently. Perhaps the most important was that on the danger to the biological diversity of hundreds of varieties of plants and crops. They argued:

Andean and Amazonian biodiversity, both domesticated and wild, is put at risk for contamination through gene flow from Terminator crops, and, as Terminator seeds would not be 100% sterile in the second generation, this risk is great. Indigenous farmers who save the seeds of contaminated varieties for replanting may find that a percentage of their seeds do not germinate, potentially translating into significant yield losses. Such contamination could cause farmers to lose trust in their own seed stock, turn their backs on traditional varieties, and increasingly depend on the purchase of Terminator varieties for harvest security so that they can guarantee at least one germination period. Similarly, the introduction of foreign genes into uncultivated varieties through gene flow from Terminator could irreversibly alter the wild varieties on which indigenous peoples have traditionally depended for important medicines and food. As a center of origin for potatoes, Peru is home to over 2,000 varieties of potatoes and is considered one of twelve megadiverse countries where 70% of the world's biodiversity resides. Biodiversity forms the basis of global food security and sovereignty for peoples and communities around the world. The spread of Terminator to indigenous agricultural systems in Peru could force indigenous farmers to abandon their traditional role as stewards of biodiversity and in doing so threaten current and future global food security. Considering that Terminator patents on potatoes have recently been claimed (Syrgenta, US Patent 6,700,039, March, 2004), the introduction of GURTs to Peru presents a high risk for irreparable contamination of this center of origin of potato.[7]

The Peruvian farmers also stressed that Terminator threatened traditional exchange of knowledge and invaluable experience among farmers:

> Traditional knowledge and innovation systems of Andean and Amazonian indigenous peoples are built around seed saving and seed exchange between plant breeders, particularly as evidenced by the extensive crop and seed exchanges at the popular weekly barter markets in the communities of Qachin, Choquecancha, Lares and Wakawasi in the district of Lares. Terminator technology would have a concrete impact on these knowledge systems by jeopardizing the availability of fertile seeds for collective exchange and breeding. As a consequence of Terminator, the very processes of adaptive interaction between man and the climatically complex Andean and Amazonian ecosystems which has allowed for the evolution and current vitality of a highly specialized body of indigenous knowledge would be paralyzed.[8]

In fact, GURTs, more popularly referred to as Terminator seeds, were also a threat to the food security of North America, Western Europe, Japan and anywhere Monsanto and its elite cartel of GMO agribusiness partners entered a market.[9] What few were aware of, however, was that the proliferation of deadly Terminator seeds might have already inadvertently been released as a result of a natural disaster.

In August 2005, two of Delta & Pine Land's greenhouses were destroyed and eleven others were damaged by a tornado. Delta & Pine Land was testing Terminator seeds in greenhouses. The company declined to inform the public whether there were Terminator tests in the houses that were destroyed or what bio-safety risks, if any, might be posed. The event showed that even seemingly secure physical containment was vulnerable. It also may have unleashed a Terminator pollution plague on the world. That would take years to determine.[10]

Selling Seeds of Destruction Everywhere
The Terminator deal closed the circle for Monsanto to emerge as the overwhelming monopolist of agricultural seeds of nearly every

variety. A year before the Delta & Pine Land bid, Monsanto had paid more than $1.4 billion for a loss-making California GMO seed giant, Seminis. Seminis, active in the patenting of GMO seeds for fruit and vegetable varieties, was the world leader in marketing vegetable and fruit plant seeds.

Seminis boasted at the time, "if you've had a salad, you've had a Seminis product."[11] At the time Monsanto took it over, the company controlled over 40% of all US vegetable seeds sold and 20% of the world market. They supplied the genetics for 55% of all lettuce on US supermarket shelves, 75% of all tomatoes and 85% of all peppers, with large shares of spinach, broccoli, cucumbers, and peas. Their seeds, primarily sold to large supermarket chains, were also widely used by conventional and organic farmers.[12]

The purchase pushed Monsanto past rival, DuPont (Pioneer Seed), to create the world's largest seed company, first in vegetables and fruits, second in agronomic crops, and the world's third largest agrochemical company. With the final acquisition of Delta & Pine Land in 2007, Monsanto was moving itself into position to hold absolute control over the majority of the planet's agricultural seeds for plants. That was not sufficient however. They were also moving into a highly controversial genetic engineering and patenting of animal seeds.

Patents on the Semen of Pigs and Bulls?

In August 2005 researchers in Germany uncovered a European patent application by Monsanto Corporation which set off new alarm bells over the scope of the attempt by private agribusiness giants to control, patent and license the entire food supply of the planet.

Monsanto had filed application for patent rights internationally with the World Intellectual Property Organization on what it claimed was its development through genetic engineering of a means to identify specific genes in pigs. Of course, the genes had come from semen provided by genetically-altered Monsanto-patented boars or male swine.[13]

Monsanto spokesperson, Chris Horner, claimed that the company merely wanted protection for its selective breeding processes, apparently a kind of eugenics for pigs, including the means to identify specific genes in pigs and use of a specialized insemination device. "We're talking about the process itself, Horner stated."[14]

The actual wording of the patent application refuted Horner's claims. In addition to seeking to patent pig breeding methods, Monsanto sought patent rights and hence, the right to collect license fees for "pig offspring produced by a method...," a "pig herd having an increased frequency of a specific ... gene...," a "pig population produced by the method...," and a "swine herd produced by a method ..." respectively.[15] If accepted, these patents would grant Monsanto intellectual property rights to particular farm animals and particular herds of livestock.

"Any pigs that would be produced using this reproductive technique would be covered by these patents," Horner admitted in a *Reuters* interview. The practices Monsanto wanted to patent involved identifying genes that result in desirable traits in swine, breeding animals to achieve those traits and using a specialized device to inseminate sows deeply in a way that uses less sperm than is typically required. "We've come up with a protocol that wraps a lot of these techniques together," said Monsanto swine molecular breeding expert Mike Lohuis.[16]

There were several techniques being used to genetically engineer animals. One method used viruses, particularly so-called retroviruses, as "vectors" to introduce new genetic material into cells because they are naturally well equipped to infiltrate them. Retroviruses are a type of virus which replicates by integrating itself into the host DNA and is then copied with the host genetic material as the cell divides.

A second method involves use of embryonic stem cells. To date however, despite many attempts to obtain ES cells from rats and farm animals, ES cells had only been isolated from some strains of mice. The technique allowed for more selective modification techniques with some control over the integration site. For example, modification can be targeted so that a transgene replaced the equivalent native gene or so that genes were "knocked out"—made ineffective by removal or disruption. A third technique was called "sperm mediated transfer." Genetically modified sperm was used as a vector for introducing foreign DNA into the egg. It had obvious attractions as artificial insemination of livestock and poultry was routine. These were the kinds of techniques being patented as fast as the GMO industry lawyers could file patent applications.[17]

1980 US Supreme Court Ruling

The Rockefeller Foundation's decades long nurturing of the field of molecular biology, its financing of the project for sequencing of genomes and the development of cloning, had led biotech giants such as Monsanto or Cargill to spend huge sums of money to genetically modify animals. The companies were focussed on one goal: patents and license rights to the results. This constituted a radical and highly controversial arena for the battle for patenting life.

The door had first been opened wide to recognition of such patents by the US Supreme Court. In 1980, the United States Supreme Court in a 5-4 decision, Diamond v. Chakrabarty, declared that "anything under the sun that is made by man" is patentable. The case concerned the patenting of genetically engineered bacteria that eat oil sludge. In 1987, the US Patent and Trademark Office issued a pronouncement of the patentability, in principle, of non-human multi-cellular organisms that were not naturally occurring. It was followed by a landmark patent on the so-called "Harvard mouse" which was engineered to be susceptible to cancer.[18]

Monsanto was not alone in attempting to control entire animal genetic seed lines. In July 2006, Cargill Corporation of Minnesota, the world's largest agriculture trading company, and one of the dominating firms in beef, pork, turkey and broiler production and processing, applied for a patent, no. US 2007/0026493 A1, with the US Patent and Trademark Office. The application was titled, "Systems and Methods for Optimizing Animal Production using Genotype Information," and the application stated its purpose was to "optimize animal production based on the animal genotype information."[19] Cargill had been engaged in a joint venture with Monsanto, Renessen Feed & Processing, near Chicago, to use advanced breeding techniques and transgenics for patented sorts of feedgrains, oilseeds and other crops.[20]

With stealth, system, and a well-supported campaign of lies and distortions, the four major GMO agribusiness giants—Monsanto, Syngenta, DuPont and Dow—were moving towards the goal once dreamed of by Henry Kissinger as ultimate control: If you control the oil, you can control nations; if you control food, you control people."

The relentless pursuit of global control over oil had been the hallmark of the Bush-Cheney Administration. Few realized that pursuit of Kissinger's second goal, control over food, was also well advanced and at a dangerous point for the future of the global population. Perhaps the most effective tool in the effort of the powerful and arrogant elites behind the spread of GMO agribusiness was their calculated cultivation of the dangerous myth that "science," in the abstract, is always "progress." This naïve popular belief in the idea of scientific progress as axiom had been one of the essential tools in the process of taking control of world food as the end of the first decade of the new century neared.

Notes

1. Monsanto Corporation, *Monsanto Company to Acquire Delta and Pine Land Company for $1.5 Billion in Cash*, Press Release, 15 August 2006, in http://monsanto.mediaroom.com/index.php?s=43&item=211.

2. Andrew Pollack, "Monsanto Buys Delta and Pine Land, Top Supplier of Cotton Seeds in US", *The New York Times*, 16 August 2006.

3. See Chapter 12, endnote 9 for details.

4. Peter Costiglio, untitled email reply to author, 12 February 2007, and 9 February 2007.

5. See Chapter 12, endnote 12.

6. Cited in Lucy Sharatt, "The Public Eye Awards 2006: Delta & Pine Land", *Ban Terminator Campaign*, http://www.evb.ch/cm_data/NOM-DELTAPINE.pdf.

7. United Nations Development Program, The Convention on Biological Diversity, Fourth meeting, Granada, 23-27 January 2006, *Potential Socio economic Impacts of Genetic Use Restriction Technologies (Gurts) on Indigenous and Local Communities, ii*, Submissions from Indigenous and local communities, Indigenous Peoples of Cusco, Peru, http://www.biodiv.org.

8. *Ibid.*

9. F. William Engdahl, "Monsanto Buys 'Terminator' Seeds Company", *Financial Sense Online*, 28 August 2006, http://www.financialsense.com/editorials/engdahl/2006/0828.html.

10. Woodrow Wilkins Jr., "D&PL Storm Losses Top $1 Million", *Delta Democrat Times*, 30 August 2005.

11. Matthew Dillon, "And We Have the Seeds: Monsanto Purchases World's Largest Vegetable Seed Company", *The Seed Alliance*, http://www.seedalliance.org/index.php?page=SeminisMonsanto, 24 January 2005.

12. *Ibid.*

13. Carey Gillam, "Crop King Monsanto Seeks Pig-Breeding Patent Clout", *Reuters*, 10 August 2005.

14. Jeff Shaw, "Monsanto Looks to Patent Pigs Breeding Methods", *New Standard*, 18 August 2005, http://newstandardnews.net.

15. *Ibid.*

16. Carey Gillam, *op. cit.*

17. Gene Watch UK, *Techniques for the Genetic Modification of Animals*, http://www.genewatch.org.

18. Max F. Rothschild, *Patenting of Genetic Innovations in Animal Breeding and Genetics,* Center for Integrated Animal Genomics, Iowa State University, Ames, Iowa, http://www.poultryscience.org/pba/1952-2003/2003/2003%20Rothschild.pdf, 2003.

19. US Patent and Trademark Office, US Patent Application Publication, *Systems and Methods for Optimizing Animal Production using Genotype Information,* Pub. No. US 2007/0026493 A1, Washington, D.C., 1 February 2007.

20. Cargill Corporation website, http://www.cargill.com/about/organization/renessen.htm.

AFTERWORD

In September 2006, the WTO published part of its ruling on the case brought to court in May 2003 by US President George W. Bush, alleging a de facto European Union moratorium on GMO. The WTO judges noted that as the European Commission had in the meantime changed its procedures to approve a number of different GMO varieties for commercial use that a moratorium or official prohibition no longer existed. Unfortunately, that was true.[1]

A Preliminary Ruling in the case had been issued by a special three-man tribunal of the World Trade Organization. The WTO decision threatened to open the world's most important region for agricultural production—the European Union—to the forced introduction of genetically-manipulated plants and food products.

The WTO case had been filed by the Government of the United States, together with Canada and Argentina—three of the world's most GMO-contaminated nations.

A WTO three-judge panel, chaired by Christian Haberli, a Swiss Agriculture Ministry bureaucrat, ruled preliminarily that the EU had applied a "de facto" moratorium on approvals of GMO products between June 1999 and August 2003, contradicting Brussels's claim that no such moratorium existed. The WTO judges argued the EU was "guilty" of not following EU rules, causing "undue delay" in following WTO obligations.[2]

The secretive WTO tribunal also ruled, according to the leaked document, that formal EU government approval to plant specific GMO plants had also been unduly delayed in the cases of 24 out of 27 specific GMO products presented to the European Commission in Brussels.

The WTO tribunal recommended that the WTO Dispute Settlement Body (DSB), the world trade policeman, call on the EU to bring its practices "into conformity with its obligations under the (WTO's) SPS Agreement." That was the notorious Sanitary and Phytosanitary escape clause of the agribusiness industry allowing it to use WTO trade supremacy to trample national rights to care for the health and safety of their citizens. Failure to comply with WTO demands could result in hundreds of millions of dollars in annual fines to the European Union.[3]

The EU Commission itself, the powerful, largely unaccountable bureaucracy in Brussels which controls the daily lives of some 470 million EU citizens in 25 member countries, was split over GMO. A Danish Agriculture Commissioner was strongly pro-GMO. The EU Environment Minister, from Greece which had a law strictly banning GMO, was strongly anti-GMO. Farmers across the EU were organizing spontaneous "GMO-free" zones and putting pressure on their politicians not to bow to WTO demands. Opinion surveys showed repeatedly than European citizens, when asked, expressed a strong negative view on GMO, often by margins of 60% or more.[4]

Geneticists Who "Play at Being God"
On April 14, 2006, at his Good Friday meditations, the highest authority of the Roman Catholic Church, German-born Pope Benedict XVI made a clear and bold declaration. The Pontiff condemned genetic scientists "who play at being God."

Addressing the recent scientific developments in the field of genetic engineering, the Pope sternly warned against their attempts to "modify the very grammar of life as planned and willed by God." He attacked the geneticists' "insane, risky and dangerous ventures which attempt to take God's place without being God." With a blistering condemnation of modern social "satanic" mores, which he said were in danger of destroying humanity, Benedict XVI then spoke of a modern "anti-Genesis," a "diabolical pride aimed at eliminating the family."

It was the strongest and most explicit condemnation yet by the Church of the practice of genetic engineering of life forms, whether plant or animal. It reinforced earlier efforts by elements of that same Church over a period of decades to resist the growing onslaughts on human reproduction financed and promoted by the circles in and around the Rockefeller Foundation (from John D. III's Population Council, to Henry Kissinger's NSSM 200, to the secret human vaccination with specially-treated Tetanus injections). With the exception of a few short media quotes, the significance of the Papal comments was buried by the major international media.[5]

At the point of writing this book, it was not clear whether or not the GMO juggernaut would be stopped globally. A new conservative Chancellor in Germany, Angela Merkel, was intent on warming chilled relations with George W. Bush's Washington. In February 2007, her Cabinet met to discuss reversing the Government's cautious GMO policies and promoting GMO as the "technology of the future." The Conservative Agriculture Minister, Horst Seehofer, advocated a decisive weakening of the previous government's Law on Gene Plants. The government of Gerhard Schroeder had approved a law stating the provision that a farmer or concern which planted GMO seeds was liable for damages to the GMO-free areas were the GMO seeds to contaminate a neighboring land. That provision, opposite to the liability law regarding GMO in the USA and Canada, had acted as a major barrier preventing the widespread proliferation of GMO in Germany and most of the European Union.[6] Yet groups of German farmers in the thousands were rapidly organizing opposition. Similar resistance was growing in Poland, Croatia, Austria, Hungary, the UK, France and across the EU.

What few realized was how very vulnerable the entire GMO mafia was to criticism.New forms of media and private communication outside the mainstream were emerging into cyberspace to communicate the experiences of farmers like Glöckner in Germany, Schmeiser in Canada, or scientists like Arpad Puzstai in Scotland, who courageously risked everything to tell the world of the risks associated with the GMO project.

Alone, the potential for exercising arbitrary political and human power in the manner in which the US and UK governments had encouraged the patenting and spread of genetically engineered plants, was grounds for organizing a global ban or moratorium on GMO plants and a permanent prohibition of any patent on living plants or animals. The fact that the grandiose claims for GMO in terms of higher yields and lower herbicide use were false to boot, added to the growing opposition to GMO.

Population reduction and genetically engineered crops were part of the same broad strategy: drastic targeted reduction of the world's population—genocide—the systematic elimination of entire population groups was the result of a wilful policy, promulgated under the name of "solving the world hunger problem."

Recalling the earlier words of Henry Kissinger is telling: "Control the oil, you control a land; control the food, you control the people …." By 2006, Washington's Bush Administration seemed well on the way to securing global control of both oil and food. What was not yet clear was whether hundreds of millions of normal, health-loving citizens would decide the issues at stake were too important to leave to such people.

F. William Engdahl
July 2007

Notes

1. World Trade Organization, *Various EC Member State Safeguard Measures Prohibiting the Import and/or Marketing of Specific Biotech Products (hereafter the "Member State Safeguard Measures")*, WT/DS291-3/R, p. 343, http://www.wto.org/english/tratop_e/dispu_e/291r_4_e.pdf.

2. F. William Engdahl, "WTO, GMO and Total Spectrum Dominance: WTO Rules Put Free-Trade of Agribusiness Above National Health Concerns", *Global Research*, 29 March 2006, http://www.globalresearch.ca.

3. European Commission, Directorate-General for Trade, Brussels, *General Overview of Active WTO Dispute Settlement Cases Involving the EC as Complainant or Defendamt and of Active Cases Under the Trade Barriers Regulation*, 23 February 2007.

4. Friends of the Earth of Europe, *What Europeans Think about GMOs*, http://www.foeeurope.org/GMOs/What_Europeans.htm.

5. Ruth Gledhill, "Pope Condemns Geneticists 'Who Play at Being God'", *The Times*, 14 April 2006.

6. BUND, Deutschnald, *Drittes Gesetz zur Änderung des Gentechnikgesetzes*, http://www.bund.net.

GLOSSARY OF ACRONYMS

ADM Archer Daniels Midland
AoA Agreement on Agriculture
ARDI United States Agricultural Reconstruction and Development
  Program for Iraq
ARS Agricultural Research Service (United States)
BBSRC Biotechnology and Biological Science Research Council
BTWC Biological and Toxic Weapons Protocol
CAFOs Concentrated Animal Feeding Operations
CAP Common Agriculture Program (European Community)
CBD Convention on Biological Diversity (United Nations)
CGIAR Consultative Group on International Agricutural Research
CIAA Office of the Coordinator of Inter-American Intelligence
  Affairs
CIMMYT International Maize and Wheat Improvement Center
CPA Coalition Provisional Authority
CFR Council on Foreign Relations
CWT Consumers for World Trade
DNA Deoxyribonucleic Acid
EPA Environmental Protection Agency (United States)

ERP People's Revolutionary Army
FAO Food and Agriculture Organization (United Nations)
FDA Food and Drug Administration (United States)
FIFRA Federal Insecticide, Fungicide and Rodenticide Act
FTC Federal Trade Commission
GAO Government Accountability Office
GATT General Agreement on Tariffs and Trade
GMO Genetically Modified Organism
GURTs Genetic Use Restriction Technologies
hCG Human Chorionic Gonadotrophin
HYV High Yield Varieties
IADP Intensive Agricultural Development Program
IBEC International Basic Economy Corporation
IGF-1 Insulin-like Growth Factor 1
IMF International Monetary Fund
IPPF International Planned Parenthood Federation
IPR Intellectual Property Right
IPRB International Program on Rice Biotechnology
IRRI International Rice Research Institute
ISAAA International Service for the Acquisition of Agri-biotech
Applications
KARI Kenya Agricultural Research Institute
KWG Kaiser Wilhelm Institute for Anthropology
LDCs Least Developing Countries
MAC Mexican Agricultural Program
NEP New Economic Plan
NIH National Institutes of Health (United States)
NSSM 200 National Security Study Memorandum 200
OECD Organization for Economic Co-operation and Development
OPEC Organization of Petroleum Exporting Countries
OSS Office of Strategic Services
P&SP Packers and Stockyards Programs
PHI Pioneer Hi-Bred International
PVP Plant Variety Protection Provision
rBGH Recombinant Bovine Growth Hormone
RNA Ribonucleic Acid

SPS Sanitary and Phytosanitary Agreement
TRIPS Agreement on Trade Related Aspects of Intellectual Property
   Rights
USAF United States Air Force
USAID United States Agency for International Development
USDA United States Department of Agriculture
WHO World Health Organization
WTO World Trade Organization
WWWC World Wide Wheat Company

GLOSSARY OF TERMS

Agent Orange Herbicide and defoliant used by the US military in its Herbicidal Warfare program during the Vietnam War (1959-1975). Name derived from the orange 55-gallon drums it was shipped in.

Agent orange was used from 1961 to 1971, it was produced for the Pentagon by Monsanto, Dow Chemical and DuPont Chemical companies, among others. A dioxin, 2,3,7,8-tetra-chlorodibenzo-para-dioxin (TCDD), is produced as a byproduct of the manufacture of 2,4,5-T, one of the two components of Agent Orange, and is present in any of the herbicides that used it. The US National Toxicology Program classified TCDD to be a human carcinogen, frequently associated with soft-tissue sarcoma, non-Hodgkin's lymphoma, Hodgkin's disease and chronic lymphocytic leukemia (CLL). 2,4,5-T has since been banned for use in the US and many other countries.

Bacteria Microscopic single-celled infectious organisms. While some types may be harmless, certain bacteria can cause diseases such as mycobacterium tuberculosis.

Bacillus thuringiensis (Bt) Soil bacterium that provides the genes for making insect-killing toxins, different forms of which are incorporated into genetically modified (GMO) crops. The adverse environmental impacts of Bt crops are now well documented in scientific literature, ranging from harm of non-target organisms to the evolution of resistance in insect pests. This makes it necessary to plant a high proportion of non-Bt crops for "resistance management." Aberrant gene expression in crops results in low-dose varieties of Bt which are ineffective in pest control and foster resistance. Cross pollination with non-GMO varieties creates Bt-weeds, and Bt-plants become volunteers (crops that can grow for seasons without any cultivation). Active Bt toxins are not biodegradable, and leak from plant roots into the soil where it accumulates over time. This critically impacts soil health and affects all other trophic levels of the ecosystem. A report that GM genes had transferred from GM pollen into microbes in the gut of bee larvae underscores the fact that Bt toxin genes, like all other GM genes, will spread out of control. (see Losey. J., et al, (1999) *Nature* 399,214).

Biodiversity Diversity of living organisms in a particular place.

Biotechnology Industrial use of biological processes, commonly used as a euphemism for the controversial term, "genetic manipulation."

Cell Smallest structural unit of all living organisms that is able to grow and reproduce independently. A cell contains a nucleus and is formed from a mass of living material surrounded by a membrane. Cells can be classified as germ-line (sperm, eggs) or somatic (body tissues).

Dioxin General term that describes a group of chemicals that are highly persistent in the environment. Class of super-toxic chemicals formed as a by-product of manufacturing, molding, or burning of organic chemicals and plastics that contain chlorine. The most toxic compound is 2,3,7,8-tetrachlorodibenzo-p-dioxin or TCDD. It is the most toxic man-made organic chemical, and is second only to radioactive waste in overall toxicity. The residents of Love Canal, Niagara Falls and Times Beach,

Missouri were forced to abandon their homes due to dioxin contamination. Dioxin is a known threat near factories that produce PVC plastic or chlorinated pesticides and herbicides and where those pesticides and herbicides have been heavily used. Dioxin became infamous during the backlash against Agent Orange during and after the Vietnam War.

Health effects such as cancer, spina bifida, autism, liver disease, endometriosis, reduced immunity, and other nerve and blood disorders have been reportedly linked to dioxin exposure. In January 2001, the US National Toxicology Program raised 2,3,7,8-TCDD cautionary level from "Reasonably Anticipated to be a Human Carcinogen" to "Known to be a Human Carcinogen."

DNA Deoxyribonucleic acid, a large, double-helix molecule that contains all the genetic information in a cell.

Eugenics Deliberate manipulation of the genetic makeup of human populations, traditionally by selective birth control, infanticide, mass murder, or genocide. Eugenicists work with genetic engineering to conduct genetic screening, *in vitro* fertilization, pre-implantation screening, germ line genetic modification, etc. Modern eugenic ideas can be traced to the British Sir Francis Galton, an amateur scientist and cousin of Charles Darwin who coined the term in 1883. In his book, *Hereditary Genius*, Glaton argued that a study of accomplished men showed they were more likely to produce intelligent and talented offspring. He concluded that it was possible to produce "a highly gifted race of men" by the process of selective breeding, which he later termed *positive* eugenics. Discouraging the reproduction of "undesirables" was subsequently termed *negative* eugenics. Eugenics was more recently associated with Nazi Germany and Hitler's Master Race purification program. After 1945 American eugenicists decided to use a less emotionally charged word, and renamed their technique, "genetics" to advance the same agenda.

Expression (Gene expression) Process through which DNA is transcribed into a message (ribonucleic acid, RNA), which becomes translated into a physical polypeptide or protein that gives structural and functional features to a cell.

Gene Biological unit of inheritance; a segment of DNA which provides the genetic information necessary to make one protein or polypeptide.

Genetic determinism Doctrine that all acts, choices and events are the inevitable consequence of antecedent sufficient causes, genetic makeup, or the sum of one's genes.

Genetic engineering Manipulation of genetic material in the laboratory. It includes isolating, copying and multiplying genes, recombining genes or DNA from different species, and transferring genes from one species to another.

Genetic map Body of information on the relative locations of genes on chromosomes. Much of the effort of the Human Genome Project is focused on mapping chromosomes.

Genetic modification Technique where individual genes can be copied and transferred to another living organism to alter its genetic make up and thus incorporate or delete specific characteristics into or from the organism. The technology is also referred to as genetic engineering, genetic manipulation or gene technology.

Gene A biological unit of inheritance; a segment of DNA that provides the genetic information necessary to make one protein.

Genome The totality of genes in an organism.

GMO Genetically Modified Organism. Any organism changed by genetic engineering, often referred to as "transgenic."

Growth hormone Protein produced by the pituitary gland that promotes growth of the whole body.

Herbicide Chemical compound used to kill weeds.

Hybrid Result of a cross between parents of different genetic types or different species.

Molecular biology Study of cellular subsystems, such as proteins and nucleic acids, including their structures, relationships to biochemical activity, repositories substances of genetic information and communication agencies. The field was initially financed and developed in the 1930's largely under financial support from the Rockefeller Foundation, who at the same time were actively funding Nazi Eugenics.

Patent Intellectual property protection which gives the owner exclusive right to exploit an invention for a fixed period of time (e.g. 17-20 years) in exchange for full disclosure of how the invention is made.

Recombinant DNA (rDNA) technology The procedure for "cutting" and "splicing" DNA to make new combinations of genes.

Recombinant Bovine Growth Hormone (rBGH or rBST (recombinant bovine somatotropin) A synthetic, genetically engineered version of BGH that is injected into a cow to artificially increase milk production. BGH is a protein hormone that occurs naturally in the pituitary glands of cattle, a factor controlling the amount of milk produced by a dairy cow.

To increase milk production, in 1995 the US Food and Drug Administration approved the sale of unlabeled milk from cows injected with Monsanto's genetically engineered bovine growth hormone, rBHG, under the brand name POSILAC®. The rBGH milk differs from natural milk chemically, nutritionally, pharmacologically and immunologically. It is contaminated with pus and antibiotics resulting from mastitis induced by the biotech hormone. Most critically, rBGH milk has high levels of abnormally potent Insulin-like Growth Factor, IGF-1; up to 10 times the levels in natural milk and over 10 times more potent. IGF-1 resists pasteurization and digestion by stomach enzymes and is well absorbed across the intestinal wall. High IGF-1 blood levels are the strongest known risk factor for prostate cancer.

Tests performed by Monsanto showed that by feeding IGF-1 at the lowest dose levels for only two weeks significant growth stimulating effects were induced in organs of adult rats. Drinking rBGH milk would thus be expected to increase blood IGF-1 levels and to increase risks of developing prostate cancer and promoting its invasiveness. Apart from prostate cancer, multiple lines of evidence have also incriminated the role of IGF-1 as risk factors for breast, colon, and childhood cancers. Because of significant risks demonstrated, rBGH milk is banned in most EU countries and Canada.

Reductionism The doctrine that complex systems can be completely understood in terms of its simplest parts. For example, an organism can be completely understood in terms of its genes, a society in terms of its individuals, and so on. The foundations of Molecular Biology are reductionist.

SPS Agreement (Agreement on the Application of Sanitary and Phytosanitary Measures) An international treaty of the World Trade Organization, negotiated during the Uruguay Round of the General Agreement on Tariffs and Trade, and entered into force with the establishment of the WTO at the beginning of 1995.

Under the SPS agreement, the World Trade Organization (WTO) sets constraints on member-states' policies relating to food safety (bacterial contaminants, pesticides, inspection and labeling) as well as animal and plant health (phytosanitary). Under the SPS agreement, national quarantine barriers can be defined by the WTO as a "technical trade barrier" used to keep out foreign competitors.

The SPS agreement gives the WTO the power to override a country's use of the *precautionary principle*—a principle that allows them to act on the side of caution if there is no scientific certainty about potential threats to human health and the environment. Even though scientists agree that it is impossible to predict all forms of damage posed by insects or pest plants, under SPS rules, the burden of proof is on nation-states to scientifically demonstrate that something is dangerous before it can be regulated.

Substantial equivalence A non-scientific term to describe new, genetically engineered crops produced by biotechnology that outwardly resemble natural corn, rice, cotton, etc. The concept of substantial equivalence has never been properly clarified. The degree of difference between a natural food and its genetically modified alternative before its "substance" ceases to be "equivalent" is not defined by legislators nor anyone else. It is exactly this vagueness that makes the concept useful to the agribusiness GMO industry.

It is the basic doctrine, signed as an Executive Order by President G.H.W. Bush in 1992 by the recommendation of Monsanto, which opened the floodgates to commercialization of GMO seeds without specific government or independent testing.

Terminator technology A "technology protection system" that renders the seeds which are saved the first generation after the first sowing sterile. Technically known as Genetic Use Restriction Technologies (GURTs), Terminator Seed Technology is patented by the US Government's USDA and Delta & Pine Land (which since August 2006 is a fully-owned Monsanto company).

There are two basic forms of Terminator or GURTs:

V-GURT produces sterile seeds that cannot be saved for future planting. The technology is restricted at the plant variety level—hence the term V-GURT.

T-GURT modifies a crop in such a way that the genetic enhancement engineered into the crop does not function until the crop plant is treated with a chemical that is sold by the biotechnology company. Farmers cannot use the enhanced trait in the crop unless they purchase the activator compound from the seed patent owner. The technology is restricted at the trait level—hence the term T-GURT.

Transgenic An organism, which due to genetic engineering in a laboratory, contains foreign DNA. Related words: transgene, transgenesis. "

Trade Related Intellectual Property Rights Agreement (TRIPS)
Treaty administered by the WTO that implements minimum standards for many forms of intellectual property (IP) regulation. It was established following the final act of the Uruguay Rounf of the General Agreement on Tariffs and Trade (GATT), in the 1994 Marrakesh Agreement.

TRIPS contains requirements that nations' laws must meet for copyright laws, including patents, and monopolies for the developers of new (GMO) plant varieties. TRIPS specifies enforcement procedures, remedies, and dispute/resolution procedures. The TRIPS agreement is to date the most comprehensive international agreement on intellectual property.

Bibliography

ADUDDELL, Robert M., and Cain, Louis P., 1981, "Public Policy Toward The Greatest Trust in the World", *Business History Review*, Summer, Harvard College, Cambridge.

ANDERSEN, Martin Edwin, 1987, "Kissinger and the Dirty War," *The Nation*, 31 October.

ASSISI FOUNDATION, BIOTHAI et al., 1998, *Biopiracy, TRIPS and the Patenting of Asia's Rice Bowl*, http://www.poptel.org.uk/panap/archives/larice.htm, May.

BLACK, Edwin, 2004, *War Against the Weak: Eugenics and America's Campaign to Create a Master Race*, Thunders' Mouth Press, New York.

BOARDMAN, Margaret Carroll, 1999, *Sowing the Seeds of the Green Revolution: The Pivotal Role Mexico and International Non-Profit Organizations Play in Making Biotechnology an Important Foreign Policy Issue for the 21st Century*, Vol. 4, Summer 99, http://www.isop.ucla.edu/profmex/volume4/3summer99/Green_Finalm.htm.

BORCK, Cornelius, 2001, *The Rockefeller Foundation's Funding for Brain Research in Germany, 1930-1950*, Rockefeller Center Archive Newsletter, Spring.

BRANFORD, Sue, 2002, "Why Argentina Can't Feed Itself," *The Ecologist*, Vol. 32 No. 8.

BRANFORD, Sue, 2004, "Argentina's Bitter Harvest", *New Scientist*, 17 April.

BROWN, Lester, 1969, Seeds of Change, Praeger, New York.

BRZEZINSKI, Zbigniew, 1970, *Between Two Ages: America's Role in the Technotronic Era*, Harper publishing house, New York.

CAETANO, Andre, 2001, "Fertility transition and the Transition of Female Sterilization in Northeastern Brazil: The Roles of Medicine and Politics", *International Union for the Scientific Study of Population*, http://www.iussp.org/Brazil2001/s10/S19_02_Cactona.pdf.

CARMAN, Judy, undated, *The Problem with the Safety of Roundup Ready Soybeans*, Flinders University, Southern Australia, http://www.biotech.info.net.

CARNEGIE, Andrew, 1889, "Wealth", *North American Review*, Vol. 148 No. 391.

CAVANAUGH-O'KEEFE, John, 2000, *The Roots of Racism and Abortion: An Exploration of Eugenics*, http://www.eugenics-watch.com/roots/index.html.

CENTRAL INTELLIGENCE AGENCY, Directorate of Intelligence, 1968, *French Actions and the Recent Gold Crisis*, Washington, D.C., 20 March.

COALITION PROVISIONAL AUTHORITY, undated, *CPA Official Documents, Order 81: Patent, Industrial Design, Undisclosed Information, Integrated Circuits and Plant Variety Law*, http://www.iraqcoalition.org/regulations/index.html#Regulations.

COHEN, Robert, 1997, *Milk—The Deadly Poison*, Argus Press, Inglewood Cliffs, NJ.

COLBY, Gerard and DENNETT, Charlotte, 1995, *Thy Will Be Done: The Conquest of the Amazon—Nelson Rockefeller and Evangelism in the Age of Oil*, HarperCollins, New York.

COMMITTEE ON RULES AND ADMINISTRATION, 1974, U.S. Senate, 93rd Congress, 2nd Session, Hearings, *The Nomination of Nelson A. Rockefeller of New York to be Vice President of the United States*, Government Printing Office,Washington, D.C.

COOK, Christopher D., 2005, "Agribusiness Eyes Iraq's Fledgling Markets", *In These Times*,15 March.

COOK, Lynn J., 2002, "Millions Served", *Forbes*, 23 December.

DEVORE, Brian, undated, "Greasing the Way for Factory Bacon, Corporate hog operations—and their lagoons—Threaten the Financial and Physical health of Family Farms", *Sustainable Farming Connection*, http://www.ibiblio.org/farming-connection.

ENGDAHL, F. William, 2004, *A Century of War: Anglo-American Oil Politics and the New World Order*, Pluto Books Ltd., London.

EVANS, Edward, 2004, "Understanding the WTO Sanitary and Phytosanitary Agreement", *EDIS document FE492*, Department of Food and Resource Economics, Florida Cooperative Extension Service, UF/IFAS, University of Florida, Gainesville, FL, August.

EWEN, Stanley and PUSZTAI, Arpad, 1999, "Effect of diets containing genetically modified potatoes expressing Galanthus nivalis lectin on rat small intestine", *The Lancet*, 16 October.

FAISSNER, Klaus, undated, *Gentechnik oder Bauern?*, http://www.arge-ja.at/gentechnik_landwirtschaft_faissner.html.

FALON, Sally, and Enig, Mary, 2002, "The Great Con-ola", *Nexus Magazine*, http://www.nexusmagazine.com, August-September.

FERRARA, Jennifer, 1998, "Revolving Doors: Monsanto and the Regulators", *The Ecologist*, September/October.

FRANK, André Gunder, 1980, *Crisis: In the World Economy*, Heinemann, London.

GM WATCH, 2004, *Wambugu Wambuzling Again: Says GM Sweet Potato a Resounding Success?*, http://www.mindfully.org/GE/2004/Wambugu-Wambuzling-Again17mar04.htm, 17 March.

GOLDBERG, Ray, 2000, *The Genetic Revolution: Transforming our Industry, Its Institutions, and Its Functions*, address to The International Food and Agribusiness Management Association (IAMA), Chicago, 26 June.

GOVERNMENT OF ARGENTINA MINISTRY OF EDUCATION, undated, *La dictadura militar en Argentina: 24 de marzo de 1976–10 de diciembre de 1983,* http://www.me.gov.ar/efeme/24demarzo/dictadura.html.

GRAIN, undated, *Iraq's new patent law: A declaration of war against farmers,* http://www.grain.org.

GRAIN, 2004, *Monsanto's Royalty Grab in Argentina,* http://www.grain.org, October.

GRANT, Madison, 1936, *The Passing of the Great Race,* Charles Scribner's Sons, New York.

GREEN, Tanya L., undated, "The Negro Project: Margaret Sanger's Genocide Project for Black Americans", *The Negro Project,* http://www.blackgenocide.org/negro.html.

HARDELL, Lennart, and ERIKSSON, Miikael, 1999, "A Case-Control Study of Non-Hodgkin Lymphoma and Exposure to Pesticides", *Cancer,* 15 March.

HARKIN, Tom, 2004, *Economic Concentration and Structural Change in the Food and Agriculture Sector,* Prepared by the Democratic Staff of the Committee on Agriculture, Nutrition, and Forestry United States Senate, 29 October.

HARR, John Ensor and JOHNSON, Peter J., 1988, *The Rockefeller Century: Three Generations of America's Greatest Family,* Scribner's, New York.

HARRIS, Robert, and PAXMAN, Jeremy, 1982, *A Higher Form of Killing,* Noonday Press, New York.

HAERLIN, Benedikt, undated, "Opinion Piece about Golden Rice", *Greenpeace International,* archive.greenpeace.org/geneng/highlights/food/benny.htm.

HIGHAM, Charles, 1983, *Trading with the Enemy: An Exposé of the Nazi-American Money Plot,1933-1947,* Delacorte, New York.

HITLER, Adolf, 1941, *Mein Kampf,* Translated by Alvin Johnson, Reynal & Hitchcock, New York.

HO, Mae-Wan, and CHING, Lim Li., 2003 "The Case for A GM-Free Sustainable World", *Independent Science Panel,* http://www.foodfirst.org/progs/global/ge/isp/ispreport.pdf, 15 June.

HOLMES, Oliver Wendell, 1926, *Carrie Buck vs. J.H. Bell.* The Supreme Court of the United States. No. 292., October Term.

HUNTINGTON, Samuel, et al., 1975, *The Crisis of Democracy: Report on the Governability of Democracies to the Trilateral Commission,* New York University Press, New York.

INTERNATIONAL MONETARY FUND, 2004, *Iraq—Letter of Intent, Memorandum of Economic and Financial Policies, and Technical Memorandum of Understanding,* Baghdad, 24 September.

ISMI, Asad, 2000, "Cry for Argentina", *Briarpatch,* September.

JACKSON, M.T., 1995, "Protecting the heritage of rice biodiversity", *GeoJournal,* Vol 335.

JAMES, Clive, 2004, "Global Status of Commercialized Biotech/GM Crops: 2004", *I S A A A,* No. 32.

JENKINS, Cate, 1990, "Criminal Investigation of Monsanto Corporation—Cover-up of Dioxin Contamination in Products—Falsification of Dioxin Health Studies", *USEPA Regulatory Development Branch,* November.

JOENSEN, Lillian and Semino, Stella, 2004, "Argentina's Torrid Love Affair with the Soybean", *Seedling,* October.

KEMP, Tage, 1932, *Report of Tage Kemp to the Rockefeller Foundation,* RF RG 1.2, Ser 713, Box 2, Folder 15, 17 November.

KENNAN, George F., 1947, "The sources of Soviet Conduct", *Foreign Affairs,* July.

KENNAN, George F., 1948, "PPS/23: Review of Current Trends in U.S. Foreign Policy", *Foreign Relations of the United States,* Volume I.

KISSINGER, Henry, 1974, *National Security Study Memorandum 200, 24 April 1974: Implications of Worldwide Population Growth for US Security and Overseas Interests, Initiating Memo.*

KIMBRELL, Andrew, 2005, "Monsanto vs. US Farmers", *The Center for Food Safety,* http://www.centerforfoodsafety.org/ Monsantovsusfarmersreport.cfm, Washington DC.

KLEIN, Naomi, 2004, "Baghdad Year Zero", *Harpers' Magazine,* September.

KRAMER, Paul, 1981, "Nelson Rockefeller and British Security Coordination", *Journal of Contemporary History,* Vol. 16.

KREBS, A. V., 1999, "Cargill & Co.'s 'Comparative Advantage in 'Free Trade", *The Agribusiness Examiner*, 26 April.

KUEHL, Stefan, 1994, *The Nazi Connection: Eugenics, American Racism, and German National Socialism*, Oxford University Press, Oxford.

KUYEK, Devlin, 2000, "ISAAA in Asia: Promoting corporate profits in the name of the poor", *GRAIN*, October, http://www.grain.org/publications/reports/isaaa.htm.

LAUGHLIN, Harry, 1914, *Report of the Committee to Study and to Report on the Best Practical Means of Cutting Off the Defective Germ-Plasm in the American Population*, Cold Spring Harbor, New York.

LAWRENCE, Geoffrey, 1987, "Agribusiness", *Capitalism and the Countryside*, Pluto Press, Sydney.

LEDERBERG, Joshua, 1987, "The Impact of Basic Research in Genetic Recombination-A Personal Account", Part I, *Annual Review of Genetics*, Vol. 21.

LEDERER, Susan E., 2002, "*Porto Ricochet*": *Joking about Germs, Cancer, and Race Extermination in the 1930s*, Oxford University Press, Oxford.

LEONARD, Thomas C., 2005, "Retrospectives: Eugenics and Economics in the Progressive Era", *Journal of Economic Perspectives*, Fall.

LEONTIEF, W., 1953, Studies in the Structure of the American Economy, International Science Press Inc., White Plains, New York.

LEONTIEF, W. and Goldberg, Ray, 1967, "The Evolution of Agribusiness", *Harvard Business School Executive Education Faculty Interviews*, http://www.exed.hbs.edu/faculty/rgoldberg.html.

LOFTUS, John and AARONS, Mark, 1994, *The Secret War Against the Jews: How Western Espionage Betrayed the Jewish People*, St. Martin's Press, New York.

LUCE, Henry, 1941, "The American Century", *Life*, 17 February.

MANN, Robert, undated, *The Selfish Commercial Gene*, "Genetically Engineered Food—Safety Problems", *PSRAST*, http://www.psrast.org/selfshgen.htm.

MARTINEZ I. PRAT, Anna-Rosa, 1998, "Genentech Preys on the Paddy Field", *Grain*, June.

MELCHER, Richard., et al., 1999, "Fields of Genes", *Business Week*, 12 April.

MEZIANI, Gundula, and Warwick, Hugh, 2002, "The Seeds of Doubt", *The Soil Association*, http://www.soilassociation.org, 17 September.

MICHIGAN DEPARTMENT OF ENVIRONMENTAL QUALITY, 2003, "Soil Movement Advisory", *Environmental Assessment Initiative*, Information Bulletin #3, June.

MILLER, James A., 1995, "Are New Vaccines Laced With Birth-Control Drugs?", *HLI Reports, Human Life International*, Vol. 13, No. 2.

MILTON, Richard, 1997, *Shattering the myths of Darwinism*, Park Street Press, Rochester, Vermont.

MONBIOT, George, 2000, "Silent Science", in *Captive State: The corporate takeover of Britain*, Pan Books, London.

NACLA, October 1975, "US Grain Arsenal", *Latin America and Empire Report*, http://www.eco.utexas.edu/facstaff/Cleaver/357Lsum_s4_NACLA_Ch2.html.

NETWORK OF CONCERNED FARMERS, 2004, *Will GM crops yield more in Australia?*, http://www.non-gm-farmers.com, 28 November.

NEW YORK COUNCIL ON FOREIGN RELATIONS, undated, *The War & Peace Studies*, http://www.cfr.org.

NORMILE, Dennis, 1999, "Rockefeller to end network after 15 years of success", *Science*, 19 November.

O'BRIEN, Thomas, 2004, *Making the Americas: U.S. Business People and Latin Americans from the Age of Revolutions to the Era of Globalization*, History Compass 2, LA 067.

ORGANIC CONSUMERS ASSOCIATION, 1999, *New Study Links Monsanto's Roundup to Cancer*, Little Marais, MN.

OSBORN, Frederick, 1937, *Summary of the Proceedings of the Conference on Eugenics in Relation to Nursing*, American Eugenics Society Papers: Conference on Eugenics in Relation to Nursing, 24 February.

OSBORN, Frederick, 1968, *The Future of Human Heredity: An Introduction to Eugenics in Modern Society*, Weybright and Talley, New York.

PARIS CLUB, 2004, *Iraq*, http://www.clubdeparis.org, 21 November.

PLAWIUK, Eugene W., 1998, "Background on Cargill Inc., the Transnational Agribusiness Giant", *Corporate Watch: GE Briefings*, http://www.archive.corporatewatch.org, November.

PERKINS, John, 2004, *The Confessions of an Economic Hit Man*, Berrett-Koehler Publishers, San Francisco.

PEW INITIATIVE ON FOOD AND BIOTECHNOLOGY, 2004, "Genetically Modified Food Crops in the United States", http://www.pewagbiotech.org, August.

POPENOE, Paul, 1933, *Applied Eugenics*, Macmillan Company, New York.

POPULATION COUNCIL, 2000, "The ICCR at 30: Pursuing New Contraceptive Leads", *Momentum: News from the Population Council*, July.

PUSZTAI, Arpad, 1999, "Why I Cannot Remain Silent", *GM-FREE magazine*, August/September.

RBGH BULLETIN, undated, *Hidden Danger in Your Milk?: Jury Verdict Overturned on Legal Technicality*, http://www.foxrBGHsuit.com.

REGAL, Philip J., 1999, *A Brief History of Biotechnology Risk: The Engineering Ideal in Biology*, Edmonds Institute, http://www.cbs.umn.edu/~pregal/GEhistory.htm. 18 July.

ROCKEFELLER III, John D., 1972, *Report of the Commission on Population Growth and the American Future*, Washington D.C, 27 March.

ROCKEFELLER III, John D., 1961, *People, Food and the Well-Being of Mankind*, Second McDougall Lecture, Food and Agriculture Organization of the United Nations.

ROCKEFELLER III, John D., 1973, *The Second American Revolution*, Harper & Row, New York.

ROWELL, Andrew, 2003, *Don't worry, it's safe to eat: The true story of GM food, BSE and Foot and Mouth*, Earthscan Publications, London.

RUDER, Thomas and KUBILLUS, Volker, 1994, *Manner hinter Hitler*, Verlag fur Polotik and Gessellshaft, Malters.

RÜDIN, Ernst, 1930, "Hereditary Transmission of Mental Diseases", *Eugenical News*, Vol. 15.

RUPKE, N. (editor), 1988, *Science, Politics and the Public Good. Essays in Honour of Margaret Gowing*, Macmillan, Basingstoke.

SCHNEIDER, W.H., 1999, *Managing Medical Research in Europe. The Role of the Rockefeller Foundation (1920s-1950s)*, CLUEB, Bologna.

SEASE, Edmund J., 2004, "History and Trends in Agricultural Biotechnology Patent Law from a Litigator's Perspective", *Seeds of Change Symposium Banquet*, http://www.ipagcon.uiuc.edu, University of Illinois, Chicago, 9 April.

SHAH, Anup, 2002, "Food Patents—Stealing Indigenous Knowledge?", *Global Issues*, http://www.globalissues.org, 26 September.

SHAND, Hope, 1994, "Patenting the Planet", *Multinational Monitor*.

SHARPLESS, John B., 1993, *The Rockefeller Foundation, the Population Council and the Groundwork for New Population Policies*, Rockefeller Archive Center Newsletter, Fall.

SHIVA, Vandana, 1998, *Biopiracy: The Plunder of Nature and Knowledge*, Green Books, Devon.

SHIVA, Vandana, 2000, "Genetically Engineered Vitamin 'A' Rice: A Blind Approach to Blindness Prevention", http://www.biotech-info.net/blind_rice.html, 14 February.

SIMON, Laurence, 1975, "The Ethics of Triage: A Perspective on the World Food Conference", *The Christian Century*, 1-8 January.

SMITH, Jeremy, 2005, "Iraq: Order 81", *The Ecologist*, February.

SMITH, J.W., 1994, *"The World's Wasted Wealth 2"*, Institute for Economic Democracy, http://www.ied.info.

SMITH, Neil, 2003, *American Empire: Roosevelt's Geographer and the Prelude to Globalization*, University of California Press, Berkeley.

SOY ONLINE SERVICE, undated, "Myths & Truths about Soy Foods", *Soy Online Service*, http://www.SoyOnlineService.co.nz.

SPITZER, Mark, 2003, "Industrial Agriculture and Corporate Power", *Global Pesticide Campaigner*, http://www.panna.org/iacp, August.

STATE OF ILLINOIS BOARD OF ADMINISTRATION, 1917, *Vol. II: Bienniel Reports of the State Charitable Institutions: October 1, 1914 to September 30, 1916*, State of Illinois.

STEVENSON, William, 1976, *A Man Called Intrepid*, Ballantine Books, New York.

STYCOS, J.M., 1954, "Female Sterilization in Puerto Rico", *Eugenics Quarterly*, no.1.

SHARATT, Lucy, undated, "The Public Eye Awards 2006: Delta & Pine Land", *Ban Terminator Campaign*, http://www.evb.ch/cm_data/NOM-DELTAPINE.pdf.

SUNSHINE PROJECT (The), "Genetic Engineering is Regularly Used to Produce Lethal Bacteria", in *Biological Weapons and Genetical Engineering*, undated, http://www.sunshine-project.org/bwintro/gebw.html.

SUURKÜLA, Jaan, 2000, "RAFI Says Terminator Seeds on Fast Track", *Problems with Genetically Engineered Food*, PSRAST, http://www.psrast.org/probobstarch.htm.

UNITED STATES SUPREME COURT, 2001, *J.E.M. Ag Supply V. Pioneer Hi-Bred*, 122 S.Ct. 593.

UNITED STATES CONGRESSIONAL RECORD, 1992, *Kissinger Associates, Scowcroft, Eagleburger, Stoga, Iraq and BNL, Statement by Representative Henry B. Gonzalez*, US House of Representatives, Page H2694, 28 April.

VENEMAN, Ann M., 2003, *Remarks by Agriculture Secretary Ann M. Veneman To the National Association of Farm Broadcasters Annual Convention,*, US Department of Agriculture, Washington D.C., Release No. 0384.03, 14 November.

WORLD DEVELOPMENT MOVEMENT, 1999, "GMOs and the WTO: Overruling the Right to Say No", http://www.wdm.org.uk, November.

WORLD TRADE ORGANIZATION, undated, *Various EC Member State Safeguard Measures Prohibiting the Import and/or Marketing of Specific Biotech Products (hereafter the "Member State Safeguard Measures")*, WT/DS291-3/R, http://www.wto.org/english/ tratop_e/dispu_e/291r_4_e.pdf.

INDEX